S0-BLM-241

HARR

Springer Series in Electrophysics
Volume 17

Edited by Manfred R. Schroeder

Springer Series in Electrophysics

Editors: Günter Ecker Walter Engl Leopold B. Felsen

Volume 1 **Structural Pattern Recognition** By T. Pavlidis

Volume 2 **Noise in Physical Systems** Editor: D. Wolf

Volume 3 **The Boundary-Layer Method in Diffraction Problems**
By V. M. Babič, N. Y. Kirpičnikova

Volume 4 **Cavitation and Inhomogeneities** in Underwater Acoustics
Editor: W. Lauterborn

Volume 5 **Very Large Scale Integration (VLSI)**
Fundamentals and Applications
Editor: D. F. Barbe 2nd Edition

Volume 6 **Parametric Electronics** An Introduction
By K.-H. Löcherer, C. D. Brandt

Volume 7 **Insulating Films on Semiconductors**
Editors: M. Schulz, G. Pensl

Volume 8 **Theoretical Fundamentals of the Ocean Acoustics**
By L. Brekhovskikh, Y. P. Lysanov

Volume 9 **Principles of Plasma Electrodynamics**
By A. F. Alexandrov, L. S. Bogdankevich, A. A. Rukhadze

Volume 10 **Ion Implantation Techniques**
Editors: H. Ryssel, H. Glawischnig

Volume 11 **Ion Implantation: Equipment and Techniques**
Editors: H. Ryssel, H. Glawischnig

Volume 12 **Fundamentals of VLSI Technology**
Editor: Y. Tarui

Volume 13 **Physics of Highly Charged Ions**
By R. K. Janev, L. P. Presnyakov, V. P. Shevelko

Volume 14 **Plasma Physics for Physicists**
By I. A. Artsimovich, R. Z. Sagdeev

Volume 15 **Relativity and Engineering**
By J. Van Bladel

Volume 16 **Elements of Random Process Theory**
By S. M. Rytov, Y. A. Kravtsov, V. I. Tatarsky

Volume 17 **Concert Hall Acoustics**
By Y. Ando

Volume 18 **Planar Circuits** for Microwaves and Lightwaves
by T. Okoshi

Volume 19 **Physics of Shock Waves in Gases and Plasmas**
By M. Liberman, A. Velikovich

Volume 20 **Kinetic Theory of Particles and Photons**
By J. Oxenius

Springer Series in Electrophysics

Yoichi Ando

Concert Hall Acoustics

With a Foreword by M. R. Schroeder

With 110 Figures

Springer-Verlag
Berlin Heidelberg New York Tokyo

Dr. Yoichi Ando, Associate Professor

Faculty of Engineering, Kobe University, Rokkodai, Nada
Kobe 657, Japan

Guest Editor:

Professor Dr. Manfred R. Schroeder

Drittes Physikalisches Institut, Universität Göttingen, Bürgerstr. 42–44
D-3400 Göttingen, Fed. Rep. of Germany

Series Editors:

Professor Dr. Günter Ecker

Ruhr-Universität Bochum, Theoretische Physik, Lehrstuhl I,
Universitätsstrasse 150, D-4630 Bochum-Querenburg, Fed. Rep. of Germany

Professor Dr. Walter Engl

Institut für Theoretische Elektrotechnik, Rhein.-Westf. Technische Hochschule,
Templergraben 55, D-5100 Aachen, Fed. Rep. of Germany

Professor Leopold B. Felsen Ph.D.

Polytechnic Institute of New York, 333 Jay Street, Brooklyn, NY 11201, USA

ISBN 3-540-13505-7 Springer-Verlag Berlin Heidelberg New York Tokyo
ISBN 0-387-13505-7 Springer-Verlag New York Heidelberg Berlin Tokyo

Library of Congress Cataloging in Publication Data. Ando, Y. (Yoichi), 1939- Concert hall acoustics. (Springer series in electrophysics ; v. 17) 1. Architectural acoustics. I. Schroeder, M. R. (Manfred Robert), 1926-. II. Title. III. Series. NA2800.A5 1985 725'.81 84-22153

This work is subject to copyright. All rights are reserved, whether the whole or part of the material is concerned, specifically those of translation, reprinting, reuse of illustrations, broadcasting, reproduction by photocopying machine or similar means, and storage in data banks. Under § 54 of the German Copyright Law where copies are made for other than private use, a fee is payable to "Verwertungs-gesellschaft Wort", Munich.

© Springer-Verlag Berlin Heidelberg 1985
Printed in Germany

The use of registered names, trademarks etc. in this publication does not imply, even in the absence of a specific statement, that such names are exempt from the relevant protective laws and regulations and therefore free for general use.

Typesetting: K+V Fotosatz GmbH, 6124 Beerfelden
Offset printing: Beltz Offsetdruck, 6944 Hemsbach/Bergstr.
Bookbinding: J. Schäffer OHG, 6718 Grünstadt.
2153/3130-543210

Foreword

NA2800
A5
1985
PHYS

When, in September 1962, Philharmonic Hall, the initial installment of New York City's *Lincoln Center for the Performing Arts,* first bowed before the world of music under the baton of its first chief tenant (Leonard Bernstein) – in the presence of the First Lady of the Republic (Jacqueline Kennedy) – acoustical expectations were soaring. But the elegant new hall on Upper Broadway, carefully designed by its acoustical consultants on the basis of encompassing data, meticulously collected by L. Beranek in concert halls around the globe left something to be desired. The high hopes held for the hall were soon dashed.

In nontechnical terms, there was a lack of "warmth" and "intimacy" (curiously, terms usually associated with another kind of human relaxation). There were also audible echoes – not from some mythical past of perfect musical balance, but of a harsher origin: the rear of the hall. The musicians, too, did not remain silent: they could not always hear each other well enough, thus making ensemble playing difficult.

The natural loquacity of the local citizens, mixed with the institution of the Manhattan cocktail circuit, made Philharmonic Hall soon the butt of many jokes (and some memorable *New Yorker* cartoons).

George Szell, the most vociferous of the leading conductors, likened the hall to a bank (on the outside) and a movie theater (on the inside). He called for more "microdiffusion" and derided the overhead acoustic panels (untranslatably) as "schwangere Frösche mit beleuchtetem Bauchnabel".

In this cacophony of complaints, Lincoln Center sought technical help from a resourceful neighbor on Lower Broadway, the American Telephone and Telegraph Company. AT&T in turn asked Bell Laboratories, who appointed the undersigned to join a committee of four "experts", chaired by the eminent physicist and former Chancellor of the University of California at Los Angeles, Vern Knudsen, to see what could be done (without building a new hall).

Bell Laboratories' charter in this rescue mission was to ascertain – by acoustic measurements – the physical facts and their potential subjective significance. As a first step, new measurement methods, based on computer-generated test tones and digital filtering, were developed, aiming for high precision in both temporal and spectral aspects of the hall's acoustic response. These measurements revealed a strong attenuation of musically important low-frequency components in the reflection from the overhead panels or "clouds". The effect was also found in model experiments performed by E. Meyer and H. Kuttruff at the University of Göttingen.

The clouds were introduced into Philharmonic Hall by the original consultants for the express purpose of interpolating "early" reflections between the direct sound and later-arriving energy. But the cloud size and shape was inadequate to diffuse low frequencies and, to compound the insufficiency, the regular, crystal-lattice like, array in which they were arranged along the ceiling led to destructive wave interference at adjacent low frequencies. (No wonder some listeners complained about a "silent movie effect" when they saw the cellists fiddle furiously but heard no corresponding sounds emerge from their instruments.)

This lack of low frequencies in the first overhead reflection revealed another low-frequency deficiency discovered by G. M. Sessler and J. E. West: a progressive attenuation of low frequencies in the *direct* sound as it grazes across the rows of seats. (This "seat effect" must exist in many other halls in which the main floor is insufficiently raked; but it is usually masked by the presence of low-frequency components in the early overhead reflections.)

As a result of these various attenuating circumstances, the low notes in the range from 100 to 250 Hz, compared to the higher frequencies, were depressed by as much as 15 dB (!) in much of the main seating area.

However, there was at least one excellent seat: "A 15" on the Second Terrace (old style, the number system has since been changed several times). Before the measurements were begun, the ushers (students of the Juilliard School of Music) had pointed this seat out as optimum in their opinion. And, lo and listen, in the measurements, too, "A 15" emerged as best by far: the gap of 15 dB between low notes and high notes was narrowed to less than 2 dB.

But there was another, less predictable, effect of the overhead panels: what reflections they interpolated arrived at a listener's two ears almost simultaneously. The subjective consequence of this lack of *lateral* reflections is a sensation of "detachment" from the sound generated on the stage, rather than a desirable feeling of *envelopment* by the music.

In order to elucidate some of the fundamental problems in concert hall acoustics the undersigned, in 1969, petitioned the German Science Foundation (DFG) to support basic research on the interplay between the physical parameters of a concert hall and its subjective quality. The work was performed at the Drittes Physikalisches Institut of the University of Göttingen with the collaboration of D. Gottlob, K. F. Siebrasse, U. Eysholdt, and Y. Ando from the University of Kobe, Japan, who joined the Göttingen group as a Humboldt Foundation Fellow.

Reliable subjective evaluations of the acoustic quality of different halls had become possible because of a new method, invented by B. S. Atal and the author of these notes, and further refined by P. Damaske, B. Wagener, and V. Mellert, that allowed the faithful reproduction, in a suitable anechoic chamber, of music played in different halls. For this purpose, two-channel *Kunstkopf* ("dummy-head") recordings were made at "strategic" locations throughout the audience area in each hall to be evaluated. The musical "input" for these tests was a recording in an anechoic environment kindly provided by the BBC Orchestra. Similar tests were also performed with *live* (but of course not completely reproducible) music by G. Plenge, H. Wilkens, P. Lehmann, and R. Wettschureck from L. Cremer's Institut für Technische Akustik at the Technical University of Berlin.

The recordings, made in some 20 different halls, were played back over two loudspeakers, after having been processed electronically in such a way that the original dummy-head ear signals were reproduced at the ears of a listener seated in a given position in front of the speakers. Listeners were thus able to make repeated *instantaneous* comparisons between different halls. In this manner, such pronounced differences as exist between the Vienna Musikvereinssaal and the Royal Festival Hall, London, become overpowering. But even very subtle acoustic distinctions are easily perceived in these direct comparisons. The wide spatial separation between different halls has finally been overcome; they are all brought together for a grand rendezvous in the same test chamber!

To avoid biasing the subjective results by misleading semantics, the use of ill-defined adjectives was religiously eschewed. This was achieved by reducing the evaluation for each pair of seats (in the same or different halls) to a simple *preference test*. Rather than describing their subjective impressions by such nebulous terms as "sweet", "dry", "cold", "warm", "rich", "narrow", "clear", "intimate", or the like, listeners had to state solely whether they preferred condition *A* or *B*. Many hundreds of such preference judgments were combined by a multidimensional scaling technique (invented by Douglas J. Carroll at Bell Laboratories) and used to construct a three-dimensional *preference space*. The two main dimensions of this spatial representation of the data could be identified as "consensus preference" and "individual differences in preference", respectively. (The third dimension was essentially "noise".) This space then represents listeners' acoustic preferences without semantic bias, while giving full weight to their different musical tastes.

Cross-correlation of the preference data with the *physical* parameters of the different halls revealed that, besides reverberation time and other well-known effects, *interaural dissimilarity* was the most important parameter governing subjective preference: The greater the dissimilarity between the two ear signals (as would obtain in old-style narrow halls with high ceilings) the greater the consensus preference, *independent of individual tastes.*

Most modern wide halls showed up with a low preference ranking, confirming the above interpretation that narrow halls are good because they provide earlier arriving, and therefore more intense, lateral sound. The importance of early lateral reflections was also stressed by A. H. Marshall and M. Barron in earlier, independent, investigations. The preponderance of lateral sound leads to a greater (preferred!) interaural dissimilarity, which in turn results in a feeling of being "enveloped" by, rather than separated from, the music.

This then was the main result of the subjective tests conducted over several years at Göttingen. They were supplemented by numerous other experiments involving sound fields created by both analog techniques and digital modification of existing concert halls. In the latter method, the impulse responses of wide halls, deficient in lateral sound, are modified on the computer by the addition of simulated lateral reflections. These modified responses are then convolved with music and subjectively evaluated. In this manner, the connection between early lateral reflections and preference was settled beyond reasonable doubt. In fact, in these tests *everything* remained unaltered, except for the addition of lateral reflections. Here we have perhaps the ultimate use of the computer and digital simulation in room acoustics!

The remaining nontrivial question, now that the causes of prior failings have been identified, is how to avoid costly mistakes in the future. Wide halls with relatively low ceilings are, unfortunately, here to stay: high building costs as well as larger (and wider!) audiences will see to that. Of course, detrimental ceiling reflections could be eliminated by sound absorption in the upper reaches of the hall. But, especially in a large modern hall whose volume has to be filled by a single instrument or voice, every "phonon" counts; there is no surplus acoustic energy available to be wasted.

The solution to this dilemma came in the 1970s: surface structures for ceiling and walls that diffuse the sound as widely as possible over the entire frequency range of interest. The design principles for such "reflection phase gratings" (as the physicist would call such structures that diffuse sound, but do not absorb it) came from the unlikely mathematical field of *number theory*.

Thus, a symbiosis of methods from a wide spectrum of scholarly disciplines – digital measurement methods, sound field reproduction and simulation, multi-dimensional scaling, and number theory – has finally elevated the art of concert hall acoustics to the level of a reliable science.

In the meantime number-theoretic sound diffusors (based on "quadratic residues" or "primitive roots") have been installed in several new halls (and numerous sound studios) with, apparently, great success. In fact, this is what one should expect, given that such diffusors break up solid specular reflections (that can also give rise to unpleasant echoes, especially, it seems, at seats occupied by sharp-eared music critics) into broad lateral patterns of *mini*-reflections that arrive at a listener's ears *laterally* rather than from straight above for the high-frequency range.

These *binaural* cares are the main theme of the present volume, augmented by detailed considerations of *monophonic* criteria, pioneered by Professor Ando, that depend critically on the temporal structure of the music for which the design is to be optimized. Basic chapters on sound transmission, acoustic simulation, and electronic techniques round out Ando's canvas of concert hall acoustics – a picture rendered not with the broad brush strokes of artistic intuition, but the pointed pencil of reproducible results.

May this book be heeded by – and prove profitable to – acousticians, architects, and musicians alike!

Göttingen and Murray Hill, February 1985 *Manfred R. Schroeder*

Preface

This book is an interdisciplinary approach to solving the perplexing problems which arise in the design of good acoustics in concert halls. It has been directed at the question of how best to identify and describe the precise qualities necessary for attaining excellent results. Through subjective preference judgments for simulated sound fields produced in an anechoic chamber, four independent factors have been identified as contributing to good acoustics. Three of them are called "temporal-monaural criteria", because they are closely associated with source signals which may be perceived by only one of two ears, i.e., total sound energy, delay of early reflections, and reverberation. The other important factor is a "spatial-binaural criterion" representing the binaural interdependence which is a measure of the spatial impression of the sound field.

It is interesting to compare the dimensions of the above criteria with those of our living universe, which has one dimension of time and three dimensions for space. Since music is a function of time both in the objective and subjective-acoustic environment, more dimensions of temporal criteria result. The ability to accurately judge the acoustical quality of alternative designs according to this theory gives the architect freedom to explore a wider variety of choices in visual expression. Also, it is hoped that the survey presented will encourage researchers, students, musicians, consultants and engineers in their further work.

Kobe, January 1985

Yoichi Ando

Acknowledgments

The experiments were carried out in the Drittes Physikalisches Institut, University of Göttingen, Federal Republic of Germany and in the Environmental Acoustics Laboratory (Professor Zyun-iti Maekawa), University of Kobe, Japan. I express my full appreciation to both institutions and their staffs. Special thanks are due to Professor Manfred R. Schroeder for his guidance and the invitations to work at the institute from 1971 till 1984. Also, I am grateful to Professor Kohsi Takano, Physiological Institute, University of Göttingen, for his aid in work on neural activities in auditory pathways, and to Dr. Peter Damaske, Dr. Dieter Gottlob, Dr. Hans Werner Strube, Dr. Herbert Alrutz and Mr. Dennis Noson for their helpful comments in preparing the manuscript. Most of the figures were traced by Mrs. Liane Liebe. I wish to thank the Alexander-von-Humboldt Foundation, Bonn, for enabling me to carry out experiments and investigations in Göttingen since 1975.

Most figures have been previously published by me or my colleagues. I express my appreciation to the authors and publishers who have granted permission for me to use already published material in this publication, duly acknowledged where appropriate. I thank Dr. Helmut Lotsch of Springer-Verlag for his arrangement to publish this book as a scientist and an editor.

Contents

1. Introduction

With the evolutionary emergence of human beings, acoustic science began with speech communication and music. Acoustics includes the environment needed to convey messages more effectively. We perceive speech signals and music by means of the sense of hearing; nevertheless, the related aspects of physical acoustics are still in question. For example, a complex extension of this problem is, what constitutes a superior sound field for concert halls? It is easy to imagine that our subjective experiences in hearing music are so strong that considerable difficulty may arise in finding a reasonable answer to this question. Obviously, it is impossible to find an answer without knowing all significant physical parameters which affect sound fields in concert halls.

In 1966, *Schroeder* [1.1] wrote an article entitled "Architectural Acoustics", which observed: "Persisting uncertainties in the acoustical design of concert halls show the need for more basic research." As an introduction to this book I should like to quote the first part of the article to outline what the problems to be solved were at that time.

"Several modern concert halls, among them La Grande Salle in Montreal, Canada, completed in 1963, and the Music Pavilion in Los Angeles, inaugurated early in 1965, have been acclaimed for their outstanding acoustical quality. Other new concert halls have been criticized for one or several acoustical deficiencies, e.g., London's Royal Festival Hall (1951), New York's Philharmonic Hall (1962), and Berlin's new Philharmonie (1963). This inconsistency in the acoustical quality of concert halls, especially large halls of modern design, attests to an insufficient understanding of the important factors required for good concert-hall acoustics. This lack of understanding is manifest to varying degrees in all three problem areas affecting concert-hall acoustics: the physical, the psychoacoustic, and the esthetic."

"The physical side of the problem is characterized by the question, 'given an enclosure with known shape and wall materials, how do sound waves travel in it?' Much uncertainty exists about important details of the reverberation process, both as a function of time (sound decay) and as a function of location and direction (sound diffusion). In fact, even measuring some of the physical parameters presents formidable obstacles."

"The psychoacoustic side of the problem can be characterized by the question 'given a known sound field, what do we *hear?*'. Unfortunately, uncertainty dominates in this field. Many basic questions relating, for example, to the subjectively perceptible differences of sound diffusion have not even been tackled, let alone answered. More complex problems, such as the identifica-

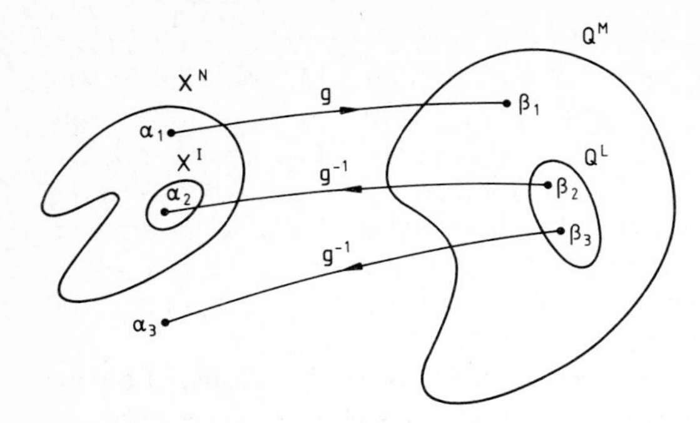

Fig. 1.1. Physically realizable space X^N and auditory subjective space Q^M. (X^I) Optimal physically realizable space; (Q^L) Optimal auditory subjective space; (g) Mapping function from the physically realizable space to the auditory subjective space, see (5.9); and (g^{-1}) Inverse function from the optimal subjective space to the physical spaces which includes a nonrealizable space. Only the realizable space may be designed

tion of the physical correlates of 'reverberance' ('liveness'), 'intimacy', 'warmth', 'immersion', and many other subjective categories, are much less well understood than has been generally assumed."

"Finally, the esthetic or 'preference' problem ('given a known sound field and complete knowledge of what is heard, what acoustical qualities do people *prefer* to hear?') raises questions that can be investigated meaningfully only on a firm basis of physical and psychoacoustic knowledge."

The problem is how to approach optimal conditions of sound fields. For simplicity, let a sound source be located on the stage and a listener be seated at a certain position in a concert hall. Let X^N be a physically realizable space with N being the independent parameters of certain boundary conditions of the sound source in the room as shown in Fig. 1.1. Also let Q^M be an auditory subjective space with M dimensions: N is less than M because of internal and external environmental factors, such as hunger, intensity of light, temperature imagination from past experience etc.

Let us suppose that a given sound field $\alpha_1 \in X^N$ corresponds to a subjective response $\beta_1 \in Q^M$ of the listener, and that these are connected by a mapping function g. This function is expected to be determined by both auditory and brain functions.

Sound fields with various combinations of objective (physical) parameters can be simulated with a digital computer. It is thereby possible to find the function connecting the physically realizable space to the auditory subjective space. If an optimal subjective space Q^L ($\subseteqq Q^M$) is found by subjective judgments in relation to objective parameters in physical space, then the corresponding sound field, say α_2, in the optimal physically realizable space X^I ($\subseteqq X^N$) can be designed in which $L \leqq M$ and $I \leqq N$.

Thus, in the optimal spaces it is expressed by

$$\{g(\alpha)\,|\,\alpha \in X^I\} = \{\beta\,|\,\beta \in Q^L\}\,. \tag{1.1}$$

Since $\alpha_3 \notin X^N$,

$$\{g^{-1}(\beta)\,|\,\beta \in Q^L\} \geqq \{\alpha\,|\,\alpha \in X^I\} \tag{1.2}$$

where $g(\phi) = \phi$ and $g^{-1}(\phi) = \phi$, ϕ being the empty set.

In the present book this problem is approached by focusing on subjective preference. At the beginning of the investigation, as a matter of fact, it was hard to accept that preference judgments alone were adequate to obtain meaningful results. Then, in an anechoic chamber, a remarkable sound field was heard, synthesized with a single reflection, with changing amplitude, time delay and direction of arrival. I concluded that the sound field would be accepted by everybody. Thereafter, I tried to simulate such sound fields with the aid of a computer with well-defined, fully independent objective parameters included in signals arriving at both ears. To minimize effects of other environmental factors, paired-comparison tests of subjective preference were performed to obtain a full range of sound fields.

The usual expressions, for example, correlation functions, Fourier transforms and convolution, are used to describe physical sound signals.

Chapter 2 is intended to explain acoustic system from the sound source to the listener's brain. Because of the importance of designing the boundary conditions of concert halls, mathematical treatment of the sound reflection and the scattering of boundaries is intentionally included in this book. But readers who are interested in the subjective attributes of sound fields, optimal design objectives, and design studies are advised to skip over to Chap. 4. In Chap. 3, to perform the subjective preference judgments and to make a search for more superior sound fields, methods of simulating the fields in concert halls with well-defined objective parameters are discussed. In Chap. 4, we consider how the law of comparative judgments enable us to get a linear psychological distance of preference between sound fields. No essential differences were obtained in the preference results of German, Japanese and Korean subjects. The sets of results agree almost completely with each other and could be predicted by an identical formula. In Chap. 5, a possible acoustic system from the sound source located on the stage to the listener's brain is discussed. Presumably, if enough were known about how the central nervous system modifies the nerve impulses from the auditory periphery, the design of auditoriums could proceed according to guidelines derived from knowledge of brain processes. An attempt is made through the auditory evoked potentials over the left and right human cerebral hemispheres to explain why the temporal and spatial objective factors independently affect the subjective preference space. The resulting optimal design objectives obtained through the temporal and spatial analyses are described. After discussing the the capability of calculating acoustical quality at any seat prior to the final architectural scheme, calculations of the scale values of preference in a concert hall are described. Chapter 6 includes some studies for designing walls, ceilings, floors, seat arrangements, the stage enclosure and whole shapes of concert halls. Chapter 7 treats acoustic measuring techniques and preference test techniques for examining sound fields in existing concert halls. An example of a diagnostic system analyzing the four objective parameters and evaluating the sound quality will be proposed.

This book may help musicians answer the question: what kind of music is the most suitable for a particular concert hall? However, it does not solve all the problems. Therefore it is hoped that some of the questions posed throughout may offer suitable lines for further research. In particular, the problems concerning the interaction between auditory and visual responses is left for future investigations.

2. Sound Transmission Systems

In this chapter, the sound transmission system from a source to a listener is discussed. Considering time-domain processing in the auditory system, the sound source is described in terms of the autocorrelation function. The medium consisting of air and boundaries is expressed by the pressure transfer function or the impulse response. The hearing system including the nervous system is briefly introduced.

2.1 Source Signals in Terms of the Autocorrelation Function

Since many kinds of music are performed in concert halls, it is therefore necessary to classify the music statistically.

In acoustics it is usual to transform a source signal $p(t)$ as a function of time into its Fourier spectrum

$$P(\omega) = \frac{1}{2\pi} \int_{-\infty}^{+\infty} p(t)\, \mathrm{e}^{-\mathrm{j}\omega t} dt\,. \tag{2.1}$$

This is a complex function of the angular frequency ω with amplitude and phase. Inversely, it may be transformed from the frequency domain into the time domain again,

$$p(t) = \int_{-\infty}^{+\infty} P(\omega)\, \mathrm{e}^{\mathrm{j}\omega t} d\omega\,. \tag{2.2}$$

The real product $P(\omega)P^*(\omega)$ gives the power density spectrum $P_\mathrm{d}(\omega)$, where $P^*(\omega)$ is the complex conjugate of $P(\omega)$.

However, in this book the source signal $p(t)$ is most often characterized by its autocorrelation function, defined by

$$\Phi(\tau) = \lim_{T\to\infty} \frac{1}{2T} \int_{-T}^{+T} p(t)\, p(t+\tau)\, dt\,, \tag{2.3}$$

where τ is the time delay and $2\,T$ is the interval of integration. Note that in a practical measurement the integration interval $2\,T$ is always limited. The autocorrelation function is equivalent to the power density spectrum $P_d(\omega)$ according to the Wiener theorem:

$$\Phi(\tau) = \int_{-\infty}^{+\infty} P_d(\omega) \cos \omega\tau \, d\omega . \tag{2.4}$$

The phase term disappears in the autocorrelation function. The autocorrelation remains useful since the ear is essentially insensitive to the absolute phase, as far as linear room acoustics is concerned.

If we set $\tau = 0$ in (2.3, 4) then

$$\Phi(0) = \lim_{T\to\infty} \frac{1}{2T} \int_{-T}^{+T} p^2(t)\, dt \qquad \text{and} \tag{2.5}$$

$$\Phi(0) = \int_{-\infty}^{+\infty} P_d(\omega)\, d\omega , \tag{2.6}$$

respectively. These equations are equivalent, and $\Phi(0)$ is the average intensity level of the signal. Since $\Phi(0)$ is the maximum value of $\Phi(\tau)$, the autocorrelation function can be normalized as

$$\phi(\tau) = \frac{\Phi(\tau)}{\Phi(0)} . \tag{2.7}$$

Let us discuss the physical meaning of (2.3). If the amplitudes of $p(t)$ and $p(t+\tau_1)$ are large and if these signals have similar repetitive features, then the integrand contributes significantly to the correlation value $\Phi(\tau_1)$.

In particular, if the values of the integrand $p(t)p(t+\tau_1)$ are mainly negative, then the total correlation $\Phi(\tau_1)$ is negative (Fig. 2.1). But at $\tau = 0$ the correlation value always reaches its positive maximum.

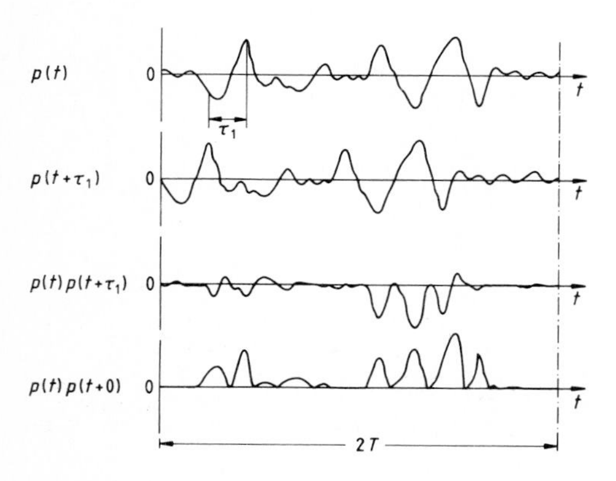

Fig. 2.1. Physical meaning of the autocorrelation function of a source signal at the origin and at a delay time τ_1.

The autocorrelation function of a pure tone is a cosine function with the period of the source signal, corresponding to an undamped oscillation. Therefore, this may be considered an example of an infinite "reverberation" system and this information is contained in the source signal itself.

Evaluating correlation functions with a finite integration time interval results in a measuring error. The variance of the autocorrelation function at large delay time τ is inverse to the interval $2T$ [2.1]:

$$\mathrm{Var}\{\Phi(\tau)\} \approx \frac{1}{2T} \int_{-\infty}^{+\infty} \Phi^2(\sigma)\, d\sigma\,. \tag{2.8}$$

As suggested by *Jansson* and *Sunberg* [2.2], who measured the spectra of music of 2 to 100 s duration, the piece of music to be analyzed should extend at least over 20 s. The longer interval $2T$ is the more convenient for acoustic design of concert halls. Therefore, the long-time autocorrelation function of 35 s duration is used, which may represent a significant factor in the subjective attributes, as discussed in Chap. 4. It is represented by

$$\Phi_p(\tau) = \lim_{T\to\infty} \frac{1}{2T} \int_{-T}^{+T} p'(t)p'(t+\tau)\, dt\,, \qquad 2T = 35\ \mathrm{s}\,, \tag{2.9}$$

where $p'(t) = p(t) * s(t)$. The function $s(t)$ was chosen as the impulse response of the A-weighting filter corresponding to ear sensitivity.

Table 2.1. Music and speech used and the effective duration of the long-time autocorrelation function

Sound source	Title	Composer	τ_e [ms] [a]
Music A	Royal Pavane	Gibbons	127 (127)
Music B	Sinfonietta, Opus 48; IV movement Allegro con brio	Malcolm Arnold	43 (35)
Music C	Symphony No. 102 in B flat major; II movement; Adagio	Haydn	(65)
Music D	Siegfried Idyll; Bar 322	Wagner	(40)
Music E	Symphony in C major, K-V no. 551, Jupiter IV movement; Molto Allegro	Mozart	38
Speech S	Poem read by a female	D. Kunikita	10 (12) [b]

[a] The effective duration of the autocorrelation function is defined by the delay τ_e at which the envelope of the normalized autocorrelation function becomes 0.1. The delay τ_e may differ slightly with different radiation characteristics of the loudspeaker used. Thus, the values in parentheses were used to calculate preferences in Sect. 4.2, while the others were used in Sect. 4.3

[b] The effective durations for different languages are considered to be similar, because the fundamental frequencies do not much differ

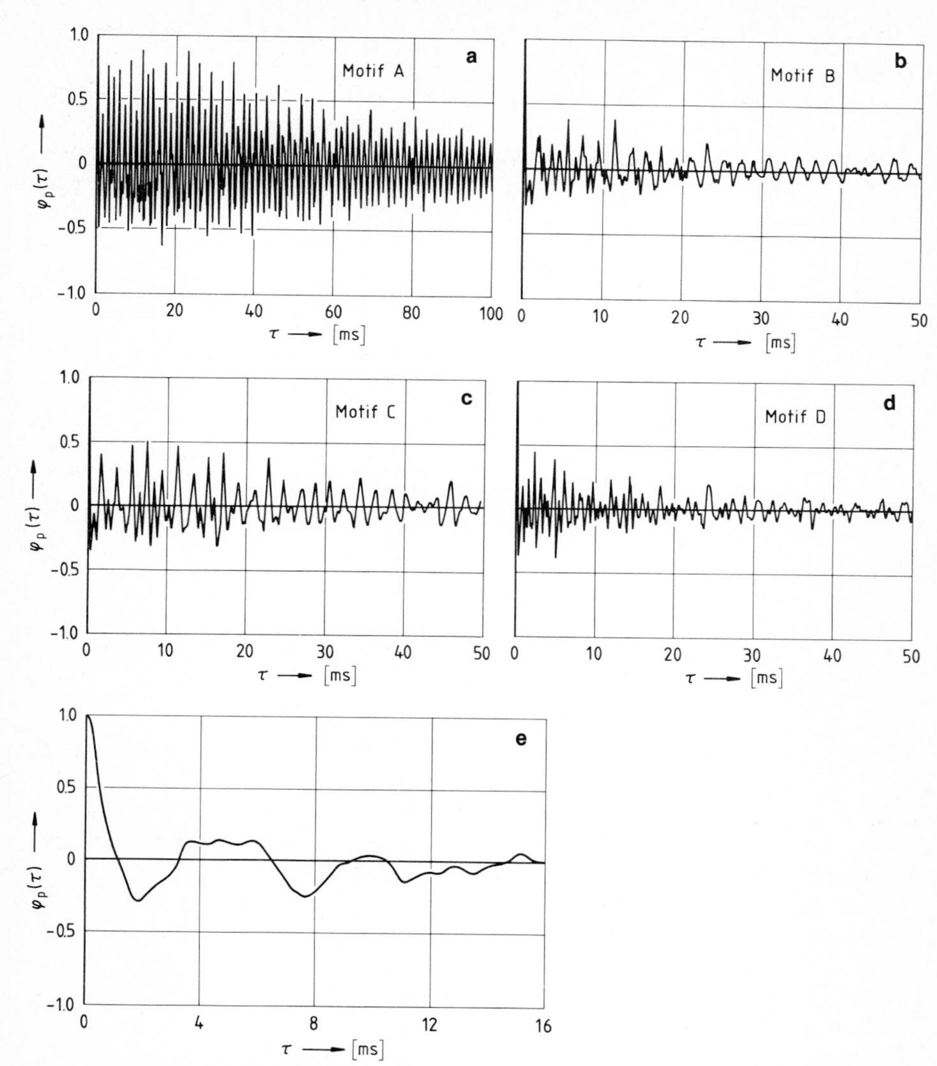

Fig. 2.2 a – e. Measured autocorrelation functions of music [2.4] and speech [2.5]. The effective duration of autocorrelation function is defined by the delay τ_e at which the envelope of the normalized autocorrelation function becomes 0.1. **(a)** Music motif A, Royal Pavane by Gibbons, $\tau_e = 127$ ms; **(b)** music motif B, Sinfonietta, Opus 48, III movement by Malcolm Arnold, $\tau_e = 35$ ms; **(c)** music motif C, Symphony No. 102 in B flat major, II movement: Adagio by Haydn, $\tau_e = 65$ ms; **(d)** music motif D, Siegfried Idyll, Bar 322 by Wagner, $\tau_e = 40$ ms; **(e)** Continuous speech, a female reading a poem, $\tau_e = 12$ ms

Samples of music measured and utilized in sound field simulations are listed in Table 2.1.

Motif A was recorded by the Philip Jones Brass Ensemble and motifs B, C, D and E were recorded by the English Chamber Orchestra in the Building Research Station anechoic chamber [2.3]. The measured, normalized autocorrelation functions are illustrated in Fig. 2.2a – d. The effective duration of the autocorrelation function is defined by the delay τ_e when the envelope of the normalized autocorrelation function becomes, and afterwards remains, smaller than 0.1 (the ten percentile delay). These values for each motif are also indicated in Table 2.1.

Extreme values are seen for music motif A by Gibbons with a slow tempo ($\tau_e = 127$ ms) and for music motif B by Arnold with a fast tempo ($\tau_e = 35$ ms). These motifs are often used in later sections. The effective duration of the other music motifs C and D are between those of A and B.

The coherence of speech signals, in general, is shorter than that of music, as shown in Fig. 2.2e. The autocorrelation function of continuous speech signals has wide peaks in the range of 3.6 – 6.3 ms which correspond roughly to the inverse of the fundamental frequency. The effective durations of speech are usually found to be about 10 – 12 ms.

2.2 Reflection from Finite Surfaces

The design of a concert hall determines its acoustic boundary conditions, e.g., the shape of a hall, its dimensions, and acoustic properties of each wall. To take into account the influence of wall properties, the transfer function for sound reflection from the walls is necessary and sufficient. Therefore, the transfer functions of several types of walls will be theoretically analyzed. But readers who are not familiar with the mathematical treatment may skip over to Sect. 2.5.

Finite rigid reflectors have often been used in concert halls. Here we discuss the sound reflection of a single rigid reflector based on Kirchhoff's well-known diffraction theory.

Let a source be located at $r_0 = (x_0, y_0, z_0)$ in front of a reflector with an arbitrary shape. Then the velocity potential of the direct sound and the reflected sound measured at $r = (x, y, z)$ may be expressed by [2.6]

$$U(r|r_0, \omega) = \frac{1}{|r - r_0|} \exp(-\mathrm{j}k|r - r_0|) + \frac{\mathrm{j}}{2\lambda} \iint_A \frac{\exp[-\mathrm{j}k(l + m)]}{lm} [\cos(n, m) - \cos(n, l)]\, ds\,, \quad (2.10)$$

where $k = \omega/c = 2\pi/\lambda$, c is the speed of sound in air, and λ the wavelength. The angles of (n, m) and (n, l) are formed between the inward normal to ds

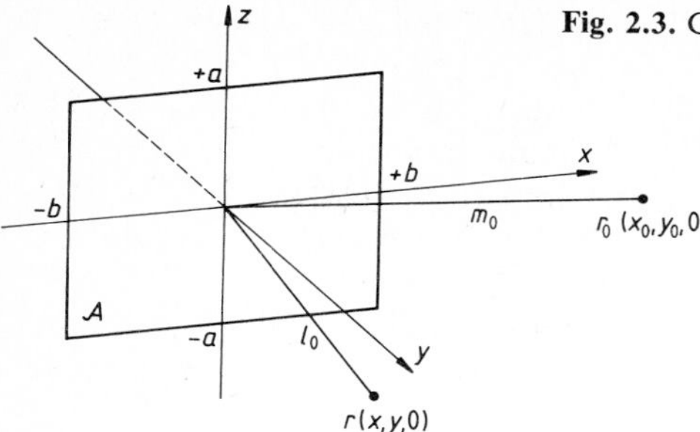

Fig. 2.3. Geometry of a rectangular reflecting surface

on the surface A and, respectively, the directions m and l, which are the distances between ds and the two points r_0 and r, respectively.

We now analyze the sound reflection of a rigid rectangular plate as shown in Fig. 2.3. Let the points r_0 and r be on the x-y plane and far from the plate, and the co-ordinates of ds be $(\eta, 0, \xi)$ on the plate, then the term $[\cos(n,m)-\cos(n,l)] = [\cos(y,m)-\cos(y,l)]$ can be approximately replaced by $[\cos\theta_i+\cos\theta_r]$, so that the reflection component U_r in (2.10) reduces to

$$U_r(r|r_0,\omega) \approx \frac{j}{2\lambda}\frac{(\cos\theta_i+\cos\theta_r)}{m_0 l_0}\exp[-jk(m_0+l_0)] \times \int_{-a}^{+a}\int_{-b}^{+b} e^{-jkf(\xi,\eta)}\,d\xi\,d\eta\,, \tag{2.11}$$

where

$$f(\xi,\eta) = (l+m)-(l_0+m_0) \quad \text{and}$$

$$l = l_0 - \frac{\eta x}{l_0} + \frac{\xi^2+\eta^2}{2l_0} - \frac{(\eta x)^2}{2l_0^3} + \frac{(\xi^2+\eta^2)\eta x}{2l_0^3} - \frac{(\xi^2+\eta^2)^2}{8l_0^3} + - \dots$$

$$m = m_0 - \frac{\eta x_0}{m_0} + \frac{\xi^2+\eta^2}{2m_0} - \frac{(\eta x_0)^2}{2m_0^3} + \frac{(\xi^2+\eta^2)\eta x_0}{2m_0^3} - \frac{(\xi^2+\eta^2)^2}{8m_0^3} + - \dots . \tag{2.12}$$

After algebraic manipulation, (2.11) may be approximately expressed by

$$U_r(r|r_0,\omega) \approx \frac{j(\cos\theta_i+\cos\theta_r)}{4m_0 l_0}\exp[-jk(m_0+l_0)].$$

$$\frac{1}{(c_0c_1)^{1/2}}\exp(jkc_2^2/4c_1)\int_{K_1^-}^{K_1^+}\exp(-j\pi v^2/2)\,dv\int_{K_2^-}^{K_2^+}\exp(-j\pi v^2/2)\,dv\,, \tag{2.13}$$

where

$$K_1^{\pm} = \pm 2\left(\frac{c_0}{\lambda}\right)^{1/2} a\,,$$

$$K_2^{\pm} = \pm 2\left(\frac{c_1}{\lambda}\right)^{1/2} b - \left(\frac{1}{\lambda c_1}\right)^{1/2} c_2\,, \quad \text{and} \tag{2.14}$$

$$c_0 = \frac{1}{2}\left(\frac{1}{m_0} + \frac{1}{l_0}\right),$$

$$c_1 = \frac{1}{2}\left(\frac{\cos^2\theta_i}{m_0} + \frac{\cos^2\theta_r}{l_0}\right), \tag{2.15}$$

$$c_2 = \sin\theta_i + \sin\theta_r\,.$$

The integral in (2.13) is known as the Fresnel integral.

The transfer function for reflection can be defined by

$$W(r|r_0,\omega) = \frac{U_r(r|r_0,\omega)}{U_i(r|r_0,\omega)}\,, \tag{2.16}$$

where the equivalent incident wave U_i is given by

$$U_i(r|r_0,\omega) = \frac{1}{m_0 + l_0}\exp[-jk(m_0 + l_0)]\,. \tag{2.17}$$

Note that the impulse response $w(t)$ is obtained by the inverse Fourier transform of $W(\omega)$ as defined by (2.2).

Thus

$$W(r|r_0,\omega) \approx \frac{j(\cos\theta_i + \cos\theta_r)}{4}\left(\frac{1}{m_0} + \frac{1}{l_0}\right)\frac{1}{(c_0 c_1)^{1/2}}\exp(jkc_2^2/4c_1)$$

$$\times[C(K_1^+) - jS(K_1^+)]\{[C(K_2^+) - jS(K_2^+)] - [C(K_2^-) - jS(K_2^-)]\}\,, \tag{2.18}$$

where

$$C(K) = \int_0^K \cos\frac{\pi v^2}{2}\,dv\,, \qquad S(K) = \int_0^K \sin\frac{\pi v^2}{2}\,dv\,.$$

When $K \to 0$,

$$C(K) \to O(K)\,, \qquad S(K) \to O(K^3)\,, \qquad \text{and when } K \to \infty,$$

$$C(\pm K) = \pm\tfrac{1}{2}\,, \qquad S(\pm K) = \pm\tfrac{1}{2}\,.$$

If $\theta_i = -\theta_r(=\theta)$, then $c_2 = 0$ and (2.18) becomes

$$W(r|r_0,\omega) \approx j2[C(K_1^+) - jS(K_1^+)][C(K_2^+) - jS(K_2^+)]\,. \tag{2.19}$$

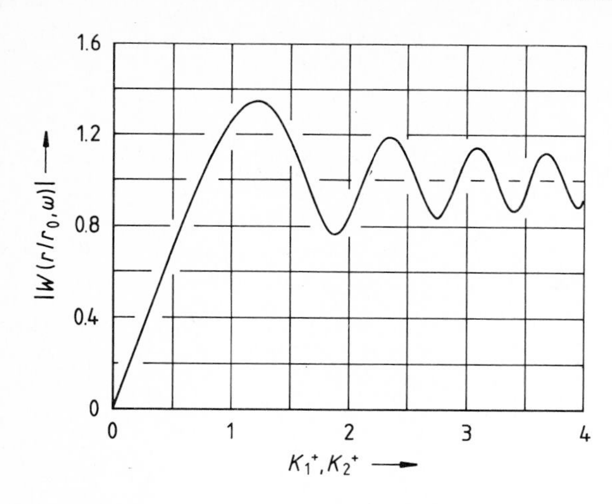

Fig. 2.4. Amplitude of the transfer function for reflection from the surface when $a \to \infty$, or $b \to \infty$

Furthermore, if a $\to \infty$, then

$$W(r|r_0, \omega)_{a\to\infty} \approx (1+\mathrm{j})[C(K_2^+) - \mathrm{j}\,S(K_2^+)]\,, \tag{2.20}$$

and if $b \to \infty$, then (2.19) becomes

$$W(r|r_0, \omega)_{b\to\infty} \approx (1+\mathrm{j})[C(K_1^+) - \mathrm{j}\,S(K_1^+)]\,, \tag{2.21}$$

where

$$K_1^+ = \left[\frac{2}{\lambda}\left(\frac{1}{m_0} + \frac{1}{l_0}\right)\right]^{1/2} a \quad \text{and}$$

$$K_2^+ = \left[\frac{2}{\lambda}\left(\frac{1}{m_0} + \frac{1}{l_0}\right)\right]^{1/2} b\cos\theta\,.$$

Equation (2.19) is particularly interesting because of the rectangular plate that may be obtained by the product of (2.20, 21) so that

$$W(r|r_0, \omega) \approx W(r|r_0, \omega)_{a\to\infty}\, W(r|r_0, \omega)_{b\to\infty}\,. \tag{2.22}$$

The reflection transfer function expressed by (2.20, 21) is therefore useful in designing the reflectors in concert halls. The absolute values of the transfer function are plotted in Fig. 2.4.

To get enough reflection in the low-frequency range, for example, both K_1^+ and K_2^+ should be greater than 0.5. If the reflectors are distributed into many parts like "clouds", there is generally an absence of low-frequency sound.

2.3 Reflection from a Periodic Structure of a Wall

Rayleigh [2.7] was the first to solve the problem of reflection from a periodic uneven surface, for which be used a cosine profile. *Bruijn* [2.8, 9] treated the problem of the reflection from a plane wave at an absorbing, periodically uneven surface of rectangular profile using *Deryugin*'s method [2.10], extending the analysis from the case of normal wave incidence and nonabsorbing structure. The method consists in solving the wave equation by separating variables both in the upper half-space and in the region containing the grooves. Using the boundary condition on the interface of two regions leads to two independent infinite systems of linear equations. These equations include the reflection amplitudes of spectral orders and the amplitudes of the groove field as unknowns.

By extending this method, a general solution for the sound reflection from a periodic surface of arbitrary profile can be obtained [2.11]. The solution can be applied to such practical purposes as the design of wall structures and the design of floor and seat combinations, as discussed in Sect. 6.2.

2.3.1 Analysis

Sound reflection from a periodically uneven surface of arbitrary profile is analyzed by dividing each groove into N rectangular regions, as shown in Fig. 2.5. The periodic surface is composed of infinitely long grooves parallel to the z axis.

Let the plane $y=0$ be chosen to coincide with the upper side of the ridges and let l be the period of the grooves. Let each range of the arbitrarily divided rectangle be

$$\nu l+a_n<x<\nu l+b_n\,, \qquad -h_n<y<-h_{n-1}\,,$$

where ν is an integer including zero and $n=1,2,\ldots,N$, as shown in Fig. 2.5. The width of the rectangle is then given by $d_n=b_n-a_n$, a_n and b_n being the lower and upper boundaries of the nth rectangular region in the x coordinate, respectively, and N being the number of these regions. Suppose that the side walls of each groove are acoustically hard. The upper side of the ridges ($y=0$) and the surfaces which coincide with the plane $y=-h_n$ are assumed to consist of locally reacting absorbing material with uniform specific acoustic admittance η_n, $-h_n$ being the depth of the nth rectangular region.

Let U_0 be the velocity potential in $y\geqq 0$ and let U_n, where $n=1,2,\ldots,N$, be the velocity potential in the nth rectangular region ($y<0$). The velocity potential must satisfy

$$\nabla^2 U_n+k^2 U_n=0\,, \qquad n=0,1,2,\ldots,N\,. \tag{2.23}$$

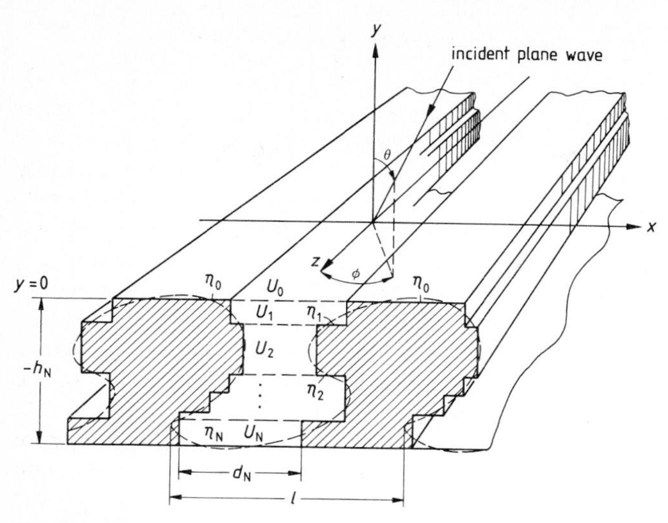

Fig. 2.5. Geometry of a periodic structure of wall [2.11]

The incident plane wave is given by

$$U_{\mathrm{i}}(x,y,z)=\exp[\mathrm{j}(\alpha_0 x+\beta_0 y+\gamma_0 z)]\ ,$$

where $\alpha_0=k\ \sin\theta\sin\phi$, $\beta_0=k\ \cos\theta$, $\gamma_0=k\ \sin\theta\cos\phi$, θ and ϕ being the angles of incidence as defined in Fig. 2.5. The time factor $\exp(\mathrm{j}\,\omega t)$ is omitted for simplicity. Because of boundary conditions, the wave equation is invariable in the z direction so that U_n may be written as

$$U_n(x,y,z)=\exp(\mathrm{j}\,\gamma_0 z)\,U_n(x,y)\ ,\qquad n=0,1,2,\ldots,N\,. \tag{2.24}$$

The incident plane wave, as a function of x and y, is given by

$$U_{\mathrm{i}}(x,y)=\exp[\mathrm{j}(\alpha_0 x+\beta_0 y)]\ .$$

Inserting (2.24) into (2.23) gives the following two-dimensional equation

$$\frac{\partial^2 U_n}{\partial x^2}+\frac{\partial^2 U_n}{\partial y^2}+\kappa^2 U_n=0\ , \tag{2.25}$$

where $\kappa^2=k^2-\gamma_0^2$.

Considering the periodicity of the uneven surface, we may write the x coordinate as $x+\nu l$, so that the range of x in $y>0$ is $-l/2<x<l/2$ and the range of x in $y<0$ is $a_n<x<b_n$ for each rectangular region. Note that the value of a_n is not always negative.

The boundary conditions to be satisfied by the equation are

i) $\dfrac{\partial U_n}{\partial x}=0,\quad$ at $x=a_n,\ x=b_n,\ -h_n<y<-h_{n-1}\ (n=1,2,\ldots,N)$;

ii) $\dfrac{\partial U_0}{\partial y}=\mathrm{j}k\eta_0 U_0,\quad$ at $y=0,\ -l/2<x<a_1$ and $b_1<x<l/2$;

iii) $\dfrac{\partial U_0}{\partial y}=\dfrac{\partial U_1}{\partial y},\quad$ at $y=0,\ a_1<x<b_1$;

iv) $U_0=U_1,\quad$ at $y=0,\ a_1<x<b_1$;

v) $\dfrac{\partial U_N}{\partial y}=\mathrm{j}k\eta_N U_N,\quad$ at $y=-h_N,\ a_N<x<b_N,\ y=-h_N$ (bottom of the groove).

The boundaries of the interface of the nth rectangular region and the $(n+1)$th region in the groove are classified into two possible profiles so that one is able to analyze an arbitrary groove profile.

Profile [a]. When $a_n \leqq a_{n+1}$ and $b_n \geqq b_{n+1}$ (Fig. 2.6a), the boundary conditions are

ai) $\dfrac{\partial U_n}{\partial y}=\mathrm{j}k\eta_n U_n,\quad$ at $y=-h_n,\ a_n<x<a_{n+1},\ b_{n+1}<x<b_n$;

aii) $\dfrac{\partial U_n}{\partial y}=\dfrac{\partial U_{n+1}}{\partial y},\quad$ at $y=-h_n,\ a_{n+1}<x<b_{n+1}$;

aiii) $U_n=U_{n+1},\quad$ at $y=-h_n,\ a_{n+1}<x<b_{n+1}$;

where $n=1,2,\ldots,N-1$.

Profile [b]. When $a_n \geqq a_{n+1}$ and $b_n \leqq b_{n+1}$ (Fig. 2.6b), the conditions are

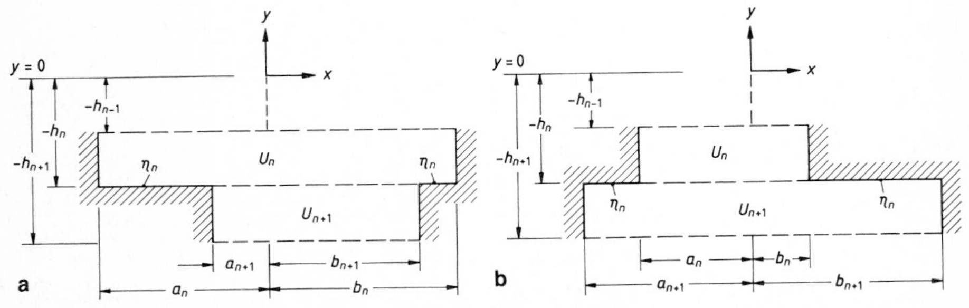

Fig. 2.6a, b. Two possible profiles of the structure [2.11]. **(a)** Profile [a]; **(b)** profile [b]

bi) $\dfrac{\partial U_{n+1}}{\partial y} = -jk\eta_n U_{n+1}$, at $y = -h_n$, $a_{n+1} < x < a_n$, $b_n < x < b_{n+1}$;

bii) $\dfrac{\partial U_n}{\partial y} = \dfrac{\partial U_{n+1}}{\partial y}$, at $y = -h_n$, $a_n < x < b_n$;

biii) $U_n = U_{n+1}$, at $y = -h_n$, $a_n < x < b_n$;

where $n = 1, 2, \ldots, N-1$. Due to the periodicity, the field in $y > 0$ can be expressed as a superposition of plane waves which either propagate along the uneven surface or are evanescent away from the surface, so that

$$
\begin{aligned}
U_0 &= \exp(j\,\alpha_0 v l)\,[U_i(x, y) + U_r(x, y)] \\
&= \exp(j\,\alpha_0 v l)\left\{\exp[j(\alpha_0 x + \beta_0 y)] + \sum_{r=-\infty}^{\infty} R_r \exp[j(\alpha_r x - \beta_r y)]\right\},
\end{aligned} \tag{2.26}
$$

where

$$
\alpha_r = \alpha_0 + 2\pi r/l,
$$

$$
\beta_r = \begin{cases} (\kappa^2 - \alpha_r^2)^{1/2} & \text{when } \kappa^2 \geqq \alpha_r^2 \\ -j(\alpha_r^2 - \kappa^2)^{1/2} & \text{when } \kappa^2 < \alpha_r^2 \end{cases}
$$

and R_r is the scattering amplitude of the rth spectral order.

In particular, the coefficient R_0 corresponds to the specular reflection component. The reflected field consists of damped and undamped outgoing waves along the y axis which are dependent on the behavior of β_r in (2.26), β_r having a purely real or purely imaginary value according to the values of kl and angles of incidence (θ, ϕ). If $\beta_r (r \neq 0)$ is real, the reflected wave of spectral order r is undamped and scattered reflection occurs. The specular reflection component always exists, because the value of β_0 is real.

The field in each rectangular region can be expressed as a superposition of waveguide modes. From the boundary condition (i) the appropriate eigenfunction is

$$
\cos\left\{m\pi\left[\left(x - \frac{a_n + b_n}{2}\right)/d_n - \frac{1}{2}\right]\right\} = \cos[m\pi(x - b_n)/d_n],
$$

where

$$
d_n = b_n - a_n \quad \text{and} \quad m = 0, 1, 2, \ldots,
$$

so that the velocity potential is expressed by

$$
\begin{aligned}
U_n = \exp(j\,\alpha_0 v l) &\sum_{m=0}^{\infty} [A_{n,m} \exp(j\,\chi_{n,m} y) + B_{n,m} \exp(-j\,\chi_{n,m} y)] \\
&\times \cos\left(m\pi \frac{(x - b_n)}{d_n}\right), \quad n = 1, 2, \ldots, N.
\end{aligned} \tag{2.27}
$$

Here

$$\chi_{n,m} = \begin{cases} (\kappa^2 - m^2\pi^2/d_n^2)^{1/2} & \text{when} \quad \kappa^2 \geqq m^2\pi^2/d_n^2 \\ -\mathrm{j}(m^2\pi^2/d_n^2 - \kappa^2)^{1/2} & \text{when} \quad \kappa^2 < m^2\pi^2/d_n^2 . \end{cases}$$

Also, $A_{n,m}$ and $B_{n,m}$ are the field amplitudes of the nth region in the groove.
For convenience, boundary condition (iii) at $y = 0$ is replaced by

$$\frac{\partial U_0}{\partial y} - \mathrm{j}k\eta_0 U_0 = \frac{\partial U_1}{\partial y} - \mathrm{j}k\eta_0 U_1 . \tag{2.28}$$

Substituting (2.26, 27) when $n = 1$ into (2.28) we obtain

$$\begin{aligned} &(\beta_0 - k\eta_0)\exp(\mathrm{j}\alpha_0 x) - \sum_{r=-\infty}^{\infty} R_r(\beta_r + k\eta_0)\exp(\mathrm{j}\alpha_r x) \\ &= \sum_{m=0}^{\infty} [(\chi_{1,m} - k\eta_0)A_{1,m} - (\chi_{1,m} + k\eta_0)B_{1,m}] \cos\left(m\pi\frac{(x-b_1)}{d_1}\right) . \end{aligned} \tag{2.29}$$

From the boundary condition (ii), the left-hand side of (2.29) is zero for $-l/2 < x < a_1$ and $b_1 < x < +l/2$. Multiply both sides of (2.29) by $\exp(-\mathrm{j}\alpha_s x)$ and integrate the left-hand side with respect to x between the limits $-l/2$ and $l/2$, also integrate the right-hand side between the limits a_1 and b_1. Then the following equation can be obtained by virtue of the orthogonality properties of exponential functions

$$R_r = \frac{\beta_0 - k\eta_0}{\beta_0 + k\eta_0}\mu_r - \frac{d_1}{l}\sum_{m=0}^{\infty} [(\chi_{1,m} - k\eta_0)A_{1,m} - (\chi_{1,m} + k\eta_0)B_{1,m}] \frac{W^*_{m,r}}{\beta_r + k\eta_0} , \tag{2.30}$$

where

$$\begin{aligned} W_{m,r} &= \frac{1}{d_1}\int_{a_1}^{b_1} \exp(\mathrm{j}\alpha_r x)\cos\left(m\pi\frac{(x-b_1)}{d_1}\right) dx \\ &= \frac{-\mathrm{j}\alpha_r[\exp(\mathrm{j}\alpha_r b_1) - (-1)^m \exp(\mathrm{j}\alpha_r a_1)]}{d_1(\alpha_r^2 - m^2\pi^2/d_1^2)} , \qquad \mu_0 = 1 \text{ and } \mu_r = 0,\ r \neq 0 . \end{aligned}$$

The asterisk above $W_{m,r}$ indicates the complex conjugate. Substituting (2.26, 27) when $n = 1$ into condition (iv) and using the orthogonality of the cosine function over the interval of the region, we obtain

$$A_{1,m} + B_{1,m} = \varepsilon_m W_{m,0} + \varepsilon_m \sum_{r=-\infty}^{\infty} W_{m,r} R_r , \tag{2.31}$$

where $\varepsilon_0 = 1$ and $\varepsilon_m = 2$, $m \neq 0$. By substituting (2.30) into (2.31), we obtain

$$\begin{aligned} A_{1,s} + B_{1,s} = \frac{2\varepsilon_s\beta_0 W_{s,0}}{\beta_0 + k\eta_0} - \varepsilon_s\frac{d_1}{l}\sum_{m=0}^{\infty} [(\chi_{1,m} - k\eta_0)A_{1,m} \\ - (\chi_{1,m} + k\eta_0)B_{1,m}] V_{m,s} , \end{aligned} \tag{2.32}$$

where

$$V_{m,s} = \sum_{r=-\infty}^{\infty} \frac{W_{s,r} W^*_{m,r}}{\beta_r + k\eta_0}, \qquad s = 0, 1, 2, \ldots .$$

We can similarly derive the following equations from the boundary conditions for the two profiles of the groove.

Profile [a]. Condition (aii) is replaced by

$$\frac{\partial U_n}{\partial y} - \mathrm{j}k\eta_n U_n = \frac{\partial U_{n+1}}{\partial y} - \mathrm{j}k\eta_n U_{n+1} . \tag{2.33}$$

Substituting (2.27) into (2.33) using the orthogonality properties of the cosine function with condition (ai) gives

$$\begin{aligned}
&(\chi_{n,s} - k\eta_n) A_{n,s} \exp(-\mathrm{j}\chi_{n,s} h_n) - (\chi_{n,s} + k\eta_n) B_{n,s} \exp(\mathrm{j}\chi_{n,s} h_n) \\
&= \frac{\varepsilon_n}{d_n} \sum_{m=0}^{\infty} [(\chi_{n+1,m} - k\eta_n) A_{n+1,m} \exp(-\mathrm{j}\chi_{n+1,m} h_n) \\
&\quad - (\chi_{n+1,m} + k\eta_n) B_{n+1,m} \exp(\mathrm{j}\chi_{n+1,m} h_n)] \, v_{m,s,n} .
\end{aligned} \tag{2.34}$$

Here

$$\begin{aligned}
v_{m,s,n} &= \int_{a_{n+1}}^{b_{n+1}} \cos\left[\frac{m\pi}{d_{n+1}}(x - b_{n+1})\right] \cos\left[\frac{s\pi}{d_n}(x - b_n)\right] dx \\
&= \left\{\sin\left[\frac{s\pi}{d_n}(b_{n+1} - b_n)\right]\right. \\
&\quad \left. - \sin\left[\frac{s\pi}{d_n}(a_{n+1} - b_n) + m\pi\right]\right\} \Big/ \left[2\pi\left(\frac{s}{d_n} - \frac{m}{d_{n+1}}\right)\right] \\
&\quad + \left\{\sin\left[\frac{s\pi}{d_n}(b_{n+1} - b_n)\right]\right. \\
&\quad \left. - \sin\left[\frac{s\pi}{d_n}(a_{n+1} - b_n) + m\pi\right]\right\} \Big/ \left[2\pi\left(\frac{s}{d_n} + \frac{m}{d_{n+1}}\right)\right],
\end{aligned}$$

$$v_{0,0,n} = d_{n+1}, \; n = 1, 2, \ldots, N-1 .$$

From condition (aiii), we also obtain

$$\begin{aligned}
&A_{n+1,s} \exp(-\mathrm{j}\chi_{n+1,s} h_n) + B_{n+1,s} \exp(\mathrm{j}\chi_{n+1,s} h_n) \\
&= \frac{\varepsilon_s}{d_{n+1}} \sum_{m=0}^{\infty} [A_{n,m} \exp(-\mathrm{j}\chi_{n,m} h_n) + B_{n,m} \exp(\mathrm{j}\chi_{n,m} h_n)] \, v_{s,m,n} , \\
&n = 1, 2, \ldots, N-1 .
\end{aligned} \tag{2.35}$$

Profile [b]. In the same way as for profile a we obtain from conditions (bi, ii)

$$(\chi_{n+1,s}+k\eta_n)A_{n+1,s}\exp(-\mathrm{j}\,\chi_{n+1,s}h_n)-(\chi_{n+1,s}-k\eta_n)B_{n+1,s}\exp(\mathrm{j}\,\chi_{n+1,s}h_n)$$

$$=\frac{\varepsilon_s}{d_{n+1}}\sum_{m=0}^{\infty}[(\chi_{n,m}+k\eta_n)A_{n,m}\exp(-\mathrm{j}\,\chi_{n,m}h_n)$$

$$-(\chi_{n,m}-k\eta_n)B_{n,m}\exp(\mathrm{j}\,\chi_{n,m}h_n)]\,w_{m,s,n}\,. \tag{2.36}$$

Here

$$w_{m,s,n}=\int_{a_n}^{b_n}\cos\left[\frac{m\pi}{d_n}(x-b_n)\right]\cos\left[\frac{s\pi}{d_{n+1}}(x-b_{n+1})\right]dx$$

$$=\left\{\sin\left[\frac{s\pi}{d_{n+1}}(b_n-b_{n+1})\right]\right.$$

$$\left.-\sin\left[\frac{s\pi}{d_{n+1}}(a_n-b_{n+1})+m\pi\right]\right\}\Big/\left[2\pi\left(\frac{s}{d_{n+1}}-\frac{m}{d_n}\right)\right]$$

$$+\left\{\sin\left[\frac{s\pi}{d_{n+1}}(b_n-b_{n+1})\right]\right.$$

$$\left.-\sin\left[\frac{s\pi}{d_n+1}(a_n-b_{n+1})-m\pi\right]\right\}\Big/\left[2\pi\left(\frac{s}{d_{n+1}}+\frac{m}{d_n}\right)\right],$$

$$w_{0,0,n}=d_n\,,\qquad n=1,2,\ldots,N-1\,.$$

Condition (biii) gives

$$A_{n,s}\exp(-\mathrm{j}\,\chi_{n,s}h_n)+B_{n,s}\exp(\mathrm{j}\,\chi_{n,s}h_n)$$

$$=\frac{\varepsilon_s}{d_n}\sum_{m=0}^{\infty}[A_{n+1,m}\exp(-\mathrm{j}\,\chi_{n+1,m}h_n)$$

$$+B_{n+1,m}\exp(\mathrm{j}\,\chi_{n+1,m}h_n)]\,w_{s,m,n}\,,\qquad n=1,2,\ldots,N-1\,. \tag{2.37}$$

In particular, from the boundary condition (v) at the bottom of the groove, we obtain the relationship between $A_{N,m}$ and $B_{N,m}$

$$B_{N,m}=\Gamma_m A_{N,m}\qquad\text{where} \tag{2.38}$$

$$\Gamma_m=\frac{(\chi_{N,m}-k\eta_N)\exp(-2\mathrm{j}\,\chi_{N,m}h_N)}{\chi_{N,m}+k\eta_N}\,.$$

Therefore, from (2.32, 38), and also the pairs (2.34, 35) or (2.36, 37) according to the profile between the nth and the $(n+1)$th rectangular regions, we may obtain an infinite system of linear equations involving $A_{n,m}$ and $B_{n,m}$

($n = 1, 2, \ldots, N-1$; $m = 0, 1, 2, \ldots$) as unknown coefficients. The coefficients $A_{N,m}$ and $B_{N,m}$ are eliminated by (2.38) and (2.34, 35) or (2.36, 37) when $n = N-1$.

The infinite set of simultaneous equations can be solved by the method of reduction, i.e., $m = 0, 1, 2, \ldots, M$ and $s = 0, 1, 2, \ldots, M$. For example, the reflection coefficients R_r can be obtained by (2.30) after getting $A_{1,m}$ and $B_{1,m}$.

The incident power across one period of the surface and per unit length in the z direction is given by

$$P_\mathrm{i} = \tfrac{1}{2}\omega \varrho l k \cos\theta$$

where ϱ is density of the medium and the reflection power is given by

$$P_\mathrm{r} = \tfrac{1}{2}\omega \varrho l \,\mathrm{Re}\left\{\sum_{r=-\infty}^{\infty} |R_r|^2 (\kappa^2 - \alpha_r^2)^{1/2}\right\}.$$

The total reflection factor is defined as

$$R_\mathrm{f} = (P_\mathrm{r}/P_\mathrm{i})^{1/2}. \tag{2.39}$$

Also, the transfer function for reflection at a space point (x, y) is defined as

$$W(x, y, \omega) = \frac{U_\mathrm{r}(x, y, \omega)}{U_\mathrm{i}(x, -y, \omega)}$$

$$= \left\{\sum_{r=-\infty}^{\infty} R_r \exp[\mathrm{j}(\alpha_r x - \beta_r y)]\right\} / \{\exp[\mathrm{j}(\alpha_0 x - \beta_0 y)]\}, \tag{2.40}$$

so that the function includes the specular reflection term R_0 and the scattered reflection terms $R_r (r \neq 0)$.

2.3.2 Numerical Calculations

Two Rectangular Grooves ($N = 2$). The solid curve shown in Fig. 2.7 exemplifies the reflecting transfer functions, $|W(\omega)|$, of profile type b, and Fig. 2.8 shows its reflection factor. The dotted curves in Fig. 2.8 represent reflection coefficients corresponding to η_1 and η_2.

As shown in Fig. 2.7, the first-order R_{-1} is undamped for $f > f_{-1}$ and the amplitude of transfer function fluctuates greatly at the higher frequencies near f_{-1}. The critical frequencies $f_{\pm r}$, $r = 1, 2, \ldots$, may be obtained by the undamped condition $\kappa^2 \geqq \alpha_r^2$ in (2.26), so that

$$f_{\pm r} = \frac{\pm cr}{l[(1 - \sin^2\theta \cos^2\phi)^{1/2} \mp \sin\theta \sin\phi]}, \tag{2.41}$$

c being the speed of sound.

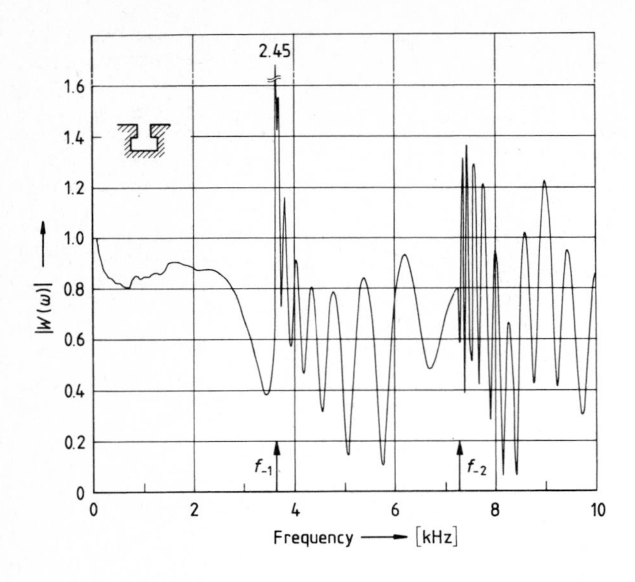

Fig. 2.7. Amplitude of the transfer function for the reflection from a wall with slit resonators at $x = 0$, $y = 50$ cm, $\phi = 90°$, $\theta = 60°$, $l = 5$ cm, $h_1 = 3$ cm, $h_2 = 6$ cm, $d_1 = 2$ cm, $d_2 = 4$ cm, $\eta_0 = 0$, ($\eta_{1,2}$ see Fig. 2.8) [2.11]

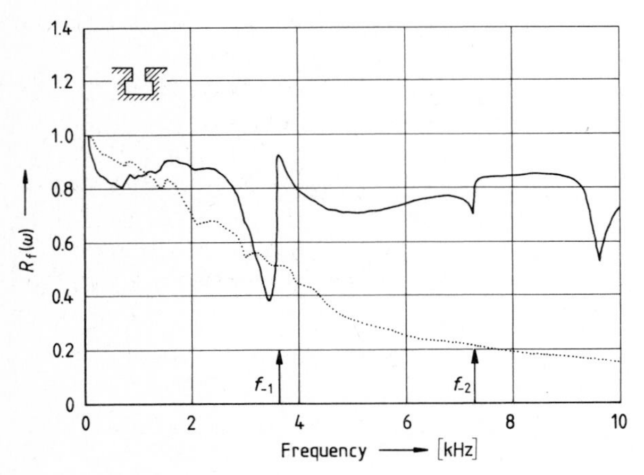

Fig. 2.8. Amplitude of the total reflection factor from the wall with slit resonators (———) [2.11]. Amplitude of the reflection from an absorbing material (bare glass fiber) which corresponds to η_1 and η_2 (··········)

The second-order R_{-2} is also undamped for $f > f_{-2}$. Only the specular reflection component is undamped for the lower frequency $f < f_{-1}$, therefore, the total reflection factor R_f is equal to the transfer function in this frequency region.

Five Rectangular Grooves ($N = 5$). As mentioned above, in this scheme it is possible to calculate the sound reflection from the surface of arbitrary profiles dividing the groove into several rectangular cross sections, which combines the two types of profile.

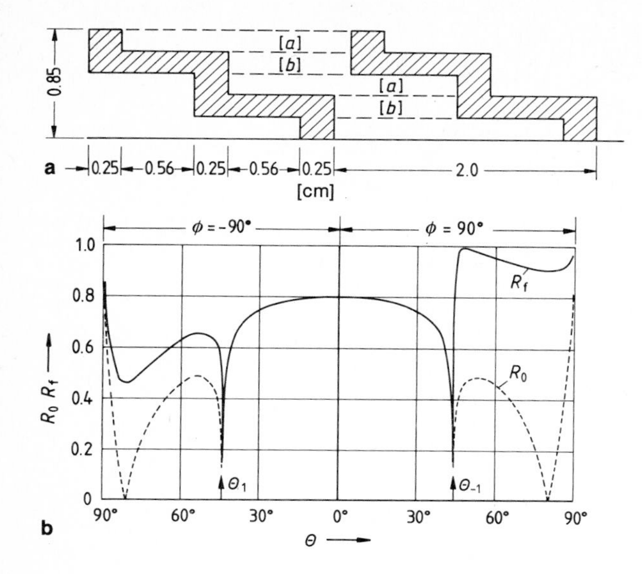

Fig. 2.9a, b. Reflection from an oblique-rib array [2.11, 12]. **(a)** Model of oblique-rib array for calculation. Combinations [a] and [b] signify the profile [a] of the structure defined in Fig. 2.6; **(b)** amplitudes of the specular reflection component R_0 and the total reflection factor R_f, 10 kHz

Figure 2.9a illustrates the calculated values of an oblique rib array. Using (2.32, 34 – 38), the linear equations involving $A_{1,m}$, $B_{1,m}, \ldots, A_{4,m}$, $B_{4,m}$ as unknowns may be solved. In this calculation, it is assumed that $\eta_0 = \eta_1 = \eta_2 = \eta_3 = \eta_4 = 0$, and $\eta_5 = 0.621 + \mathrm{j}0.190$, which corresponds to the measured value of a 10 mm thick glass fiber layer at 10 kHz [2.12]. The calculated results are shown in Fig. 2.9b as a function of the angle of incidence θ. The reflection factor is asymmetrical with the angle of incidence θ so that a large difference between the values of $\phi = 90°$ and $-90°$ is found. Unwanted reflections from a certain angle of incidence upon the wall, therefore, can be suppressed.

2.4 Scattering by Diffusing Walls

In order to design highly sound-diffusing walls over a wide frequency range, *Schroeder* [2.13] developed a special surface based on the quadratic-residue sequences of elementary number theory investigated by Lengendre and Gauss. The diffuse surface has wells of several different depths with a period N, Sect. 6.1. Here we derive a general solution for a diffusing wall with arbitrary width wells and an admittance at the bottom of each well as shown in Fig. 2.10. It may be obtained by applying the analysis described in the previous section, because the structure is periodic with a period l. For the special case of $d_i =$ const and when the admittance at the bottom of each wall is zero, a solution has already been given by *Strube* [2.14, 15]. Some of his numerical results are given in Sect. 6.1.

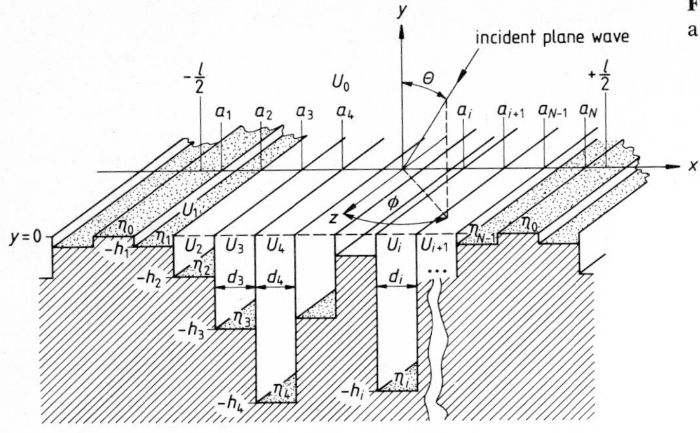

Fig. 2.10. Geometry of a periodic diffuser

The boundary conditions to be satisfied by (2.25) are

i′) $\dfrac{\partial U_\mathrm{i}}{\partial x} = 0, \quad$ at $x = a_i$ and $x = a_{i+1}$, $-h_i<y<0$; $\quad i = 1, 2, \ldots, N-1$;

ii′) $\dfrac{\partial U_0}{\partial y} = \mathrm{j}k\eta_0 U_0, \quad$ at $y = 0$, $-l/2<x<a_1$ and $a_N<x<+l/2$;

iii′) $\dfrac{\partial U_0}{\partial y} = \sum\limits_{i=1}^{N-1} \Delta_i(x)\dfrac{\partial U_i}{\partial y}, \quad$ at $y = 0$;

iv′) $U_0 = \sum\limits_{i=1}^{N-1} \Delta_i(x)\, U_i, \quad$ at $y = 0$;

v′) $\dfrac{\partial U_i}{\partial y} = \mathrm{j}k\eta_i U_i, \quad$ at $y = -h_i$, $a_i<x<a_{i+1}$, $\quad i = 1, 2, \ldots, N-1$ (bottom of each groove),

where

$$\Delta_i(x) = \begin{cases} 1, & a_i<x<a_{i+1}\,, \\ 0, & \text{elsewhere}\,. \end{cases}$$

From boundary condition (i′) the eigenfunction in each groove labeled by i is given by

$$\cos\left[m\pi(x-a_{i+1})/d_i\right]\,,$$

where $d_i = a_{i+1} - a_i$, $i = 1, 2, \ldots, N-1$ and $m = 0, 1, 2, \ldots$, so that the velocity potential is expressed by

$$U_i = \exp(\mathrm{j}\,\alpha_0 v l) \sum_{m=0}^{\infty} [A_{i,m} \exp(\mathrm{j}\,\chi_{i,m} y) + B_{i,m} \exp(-\mathrm{j}\,\chi_{i,m} y)]$$

$$\times \cos\left(m\pi \frac{(x-a_{i+1})}{d_i}\right), \qquad i = 1, 2, \ldots, N-1 . \tag{2.42}$$

The boundary condition (iii′) at $y = 0$ may be replaced by

$$\frac{\partial U_0}{\partial y} - \mathrm{j}k\eta_0 U_0 = \sum_{i=1}^{N-1} \Delta_i(x) \left(\frac{\partial U_i}{\partial y} - \mathrm{j}k\eta_0 U_i\right), \; a_i < x < a_{i+1}, \tag{2.43}$$

$i = 1, 2, \ldots, N-1$.

Substituting (2.26, 42) into the above equation for each groove $a_i < x < a_{i+1}$ at $y = 0$ gives

$$(\beta_0 - k\eta_0) \exp(\mathrm{j}\,\alpha_0 x) - \sum_{r=-\infty}^{\infty} R_r(\beta_r + k\eta_0) \exp(\mathrm{j}\,\alpha_r x)$$

$$= \sum_{m=0}^{\infty} \sum_{i=1}^{N-1} \Delta_i(x) [(\chi_{i,m} - k\eta_0) A_{i,m} - (\chi_{i,m} + k\eta_0) B_{i,m}]$$

$$\times \cos\left(m\pi \frac{(x-a_{i+1})}{d_i}\right), \qquad i = 1, 2, \ldots, N-1 . \tag{2.44}$$

Note from the boundary condition (ii′) that the left-hand side of (2.44) is zero for $-l/2 < x < a_1$ and $a_N < x < +l/2$ at $y = 0$.

Multiply both sides of (2.44) by $\exp(-\mathrm{j}\,\alpha_s x)$ and integrate the left-hand side with respect to x between the limit $-l/2$ and $+l/2$, and integrate the right-hand side between the limits a_1 and a_N. Then, the following expression is obtained by the orthogonality properties of the exponential function:

$$R_r = \frac{\beta_0 - k\eta_0}{\beta_0 + k\eta_0} \mu_r - \frac{1}{l} \sum_{i=1}^{N-1} \sum_{m=0}^{\infty} A_{i,m} [(\chi_{i,m} - k\eta_0)$$

$$- (\chi_{i,m} + k\eta_0) \Gamma_{i,m}] \frac{W^*_{i,m,r}}{\beta_r + k\eta_0}, \tag{2.45}$$

where

$$W_{i,m,r} = \int_{a_i}^{a_{i+1}} \exp(\mathrm{j}\,\alpha_r x) \cos\left(m\pi \frac{(x-a_{i+1})}{d_i}\right) dx$$

$$= \frac{-\mathrm{j}\,\alpha_r [\exp(\mathrm{j}\,\alpha_r a_{i+1}) - (-1)^m \exp(\mathrm{j}\,\alpha_r a_i)]}{(\alpha_r^2 - m^2\pi^2/d_i^2)},$$

$\mu_0 = 1$ and $\mu_r = 0, \quad r \neq 0$, and

$$\Gamma_{i,m} = \frac{B_{i,m}}{A_{i,m}} = \frac{\chi_{i,m} - k\eta_i}{\chi_{i,m} + k\eta_i} \exp(-2\mathrm{j}\,\chi_{i,m} h_i) .$$

Substituting (2.26, 42) into condition (iv′) and using the orthogonality of the cosine function over the interval of the region $a_1 < x < a_N$ at $y = 0$ gives

$$A_{i,m}(1+\Gamma_{i,m}) = \frac{\varepsilon_m}{d_i} W_{i,m,0} + \frac{\varepsilon_m}{d_i} \sum_{r=-\infty}^{\infty} W_{i,m,r} R_r, \qquad i = 1,2,\ldots,N-1,$$
$$\varepsilon_0 = 1 \quad \text{and} \quad \varepsilon_m = 2, \quad m \neq 0. \tag{2.46}$$

Further, substituting (2.45) into (2.46), we obtain

$$A_{j,s}(1+\Gamma_{j,s}) = \frac{2\varepsilon_s \beta_0 W_{j,s,0}}{d_j(\beta_0 + k\eta_0)} - \frac{\varepsilon_s}{d_j}\frac{1}{l}\sum_{i=1}^{N-1}\sum_{m=0}^{\infty} A_{i,m}[(\chi_{i,m} - k\eta_0)$$
$$-(\chi_{i,m} + k\eta_0)\Gamma_{i,m}] V_{i,j,m,s}, \qquad \text{where} \tag{2.47}$$
$$V_{i,j,m,s} = \sum_{r=-\infty}^{\infty} \frac{W_{j,s,r} W^*_{i,m,r}}{\beta_r + k\eta_0}, \qquad j = 1,2,\ldots,N-1, \qquad s = 1,2,\ldots .$$

Similarly to the method described in the previous section, we can obtain the scattering amplitudes R_r for a more general profile of walls. For example, calculations of the scattering characteristics of the periodic diffuser shown in Fig. 2.11 have been carried out. The results for these three angles of incidence (θ) are shown in Fig. 2.12 [2.16]. In addition to these results, the transfer function for scattered reflection at the listener's position, which is required to simulate the sound fields, may be calculated.

The analyses of previous and present sections and their combination may help to design a wall full of variety. Further results of calculations are shown in [2.11, 16].

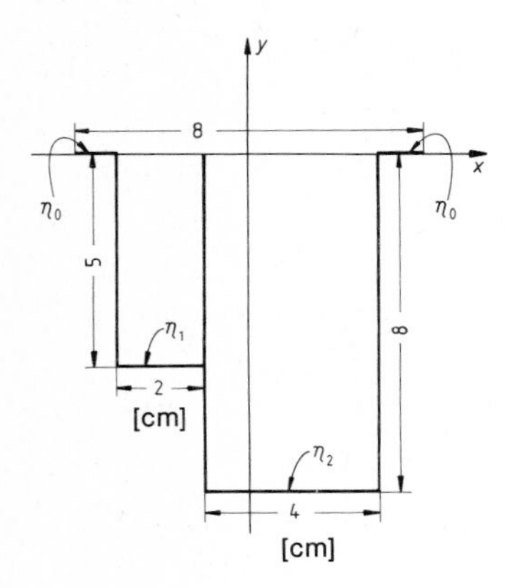

Fig. 2.11. Model of the periodic diffuser for calculation $\eta_0 = 0.508 + j0.604$, $\eta_1 = 0$, $\eta_2 = 0.219 + j0.306$

Fig. 2.12 a – c. Calculated reflection pattern of the periodic diffuser, $\phi = 90°$, 10 kHz [2.16]: **(a)** $\theta = 5.7°$; **(b)** $\theta = 45.8°$; **(c)** $\theta = 80.2°$, unit: 10 dB

a

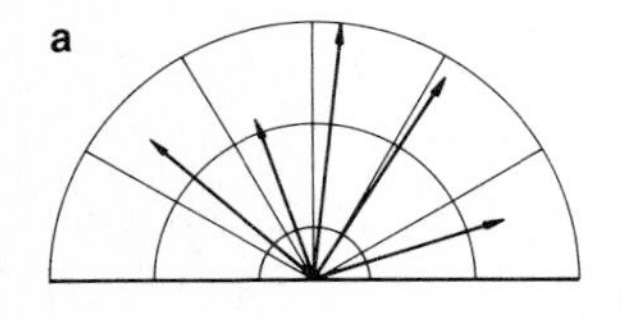

b

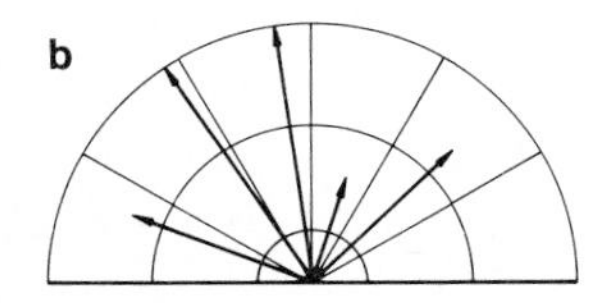

c

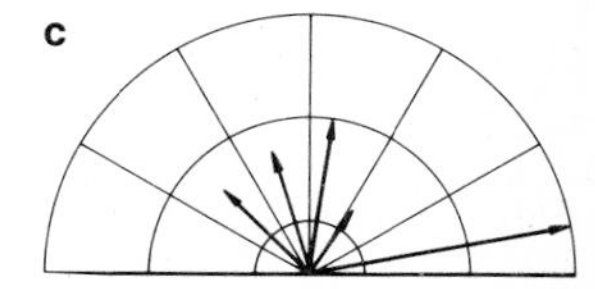

2.5 Physical Hearing System

It is quite clear that we cannot discuss concert-hall acoustics without knowing the auditory system interposed between the sound field and the brain. This system will be briefly described, before going on to discuss subjective preference judgment.

2.5.1 Head, Pinna and External Auditory Canal

The acoustic environment is perceived by the ears, in which a sound signal is given by the time sequence described in Sect. 2.1. A three-dimensional space is also perceived by the ears, mainly because the head-related transfer function $H(r|r_0, \omega)$ between a source point and the two ear entrances has directional qualities from the shapes of the head and pinna systems. The directional information is contained in the head-related transfer function, Sect. 3.2.

Figure 2.13 shows the examples of amplitude and phase of the head-related transfer function $H(\xi, \eta, \omega)$ as parameters of angle of incidence ξ [2.18] which are measured by the single-pulse method described in Sect. 7.1. The angle $\xi = 0°$ corresponds to the frontal direction and $\xi = 90°$ to the lateral direction towards the side of the ear being examined. An average group delay in the phase was eliminated.

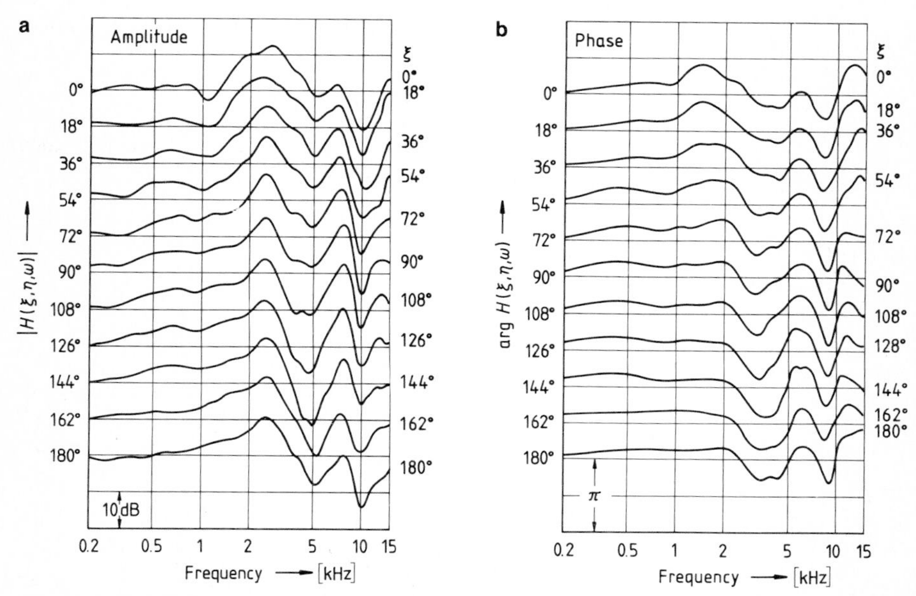

Fig. 2.13 a, b. Transfer functions from a free field to the ear-canal entrance as a parameter of the horizontal angle ξ [2.18]. **(a)** Amplitudes; **(b)** phases

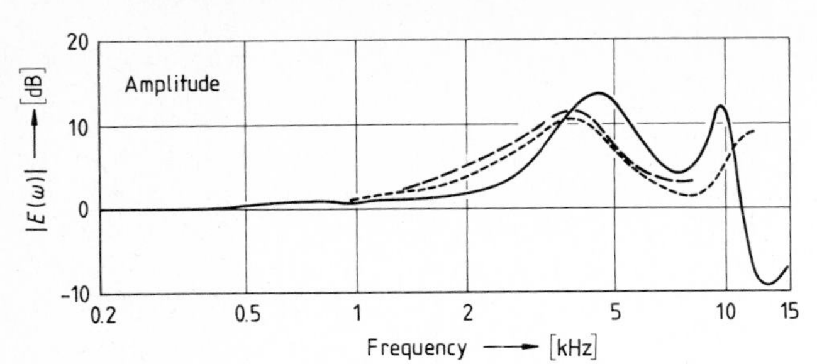

Fig. 2.14

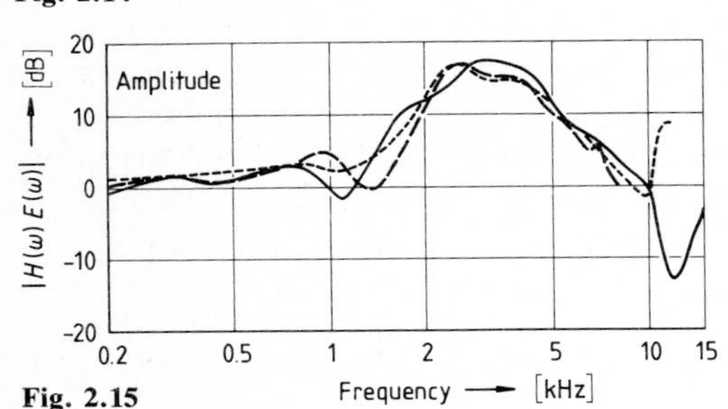

Fig. 2.15

Fig. 2.14. Transfer functions of the ear canal: (– – –) from [2.19]; (-----) from [2.20]; and (———) from [2.18]

Fig. 2.15. Transfer functions from a sound source in front of the listener to the eardrum: (– – –) from [2.19]; (-------) from [2.20]; and (———) from [2.18]

Since the diameter is small enough compared with the wavelength (8.0 kHz), the transfer function of the external canal $E(\omega)$ is independent of the directions in which sound is incident on the human head for the audio-frequency range. Therefore, interactions between the sound field in the external canal and that of the pinna are insignificant. The transfer function from the free field to the eardrum can be obtained by multiplying the following two functions: (1) the sound source in the free field to the ear-canal entrance – $H(\xi, \eta, \omega)$ – and (2) from the ear-canal entrance to the eardrum – $E(\omega)$. Measured absolute values of $E(\omega)$ are shown in Fig. 2.14, where the variations in the curves obtained by different investigations are caused by the different definitions of the ear-canal entrance point. A typical example of transfer functions from a sound source in front of the listener to the eardrum is shown in Fig. 2.15. This corresponds to direct sound when the subject is listening to a performer. The transfer functions obtained from these three reports are not significantly different for frequencies up to 10 kHz.

2.5.2 Eardrum and Bone Chain

Behind the eardrum are the tympanic cavities containing the three auditory ossicles, the malleus, incus and stapes. This area is called the middle ear (Fig. 2.16). The sound pressure striking the eardrum is transduced into vibration. The middle ear ossicles transmit the vibration to the cochlea. The vibration pattern of the human eardrum was first measured by *Békésy* [2.22] by making a point-by-point examination with a capacitive probe. Later, *Tonndorf* and

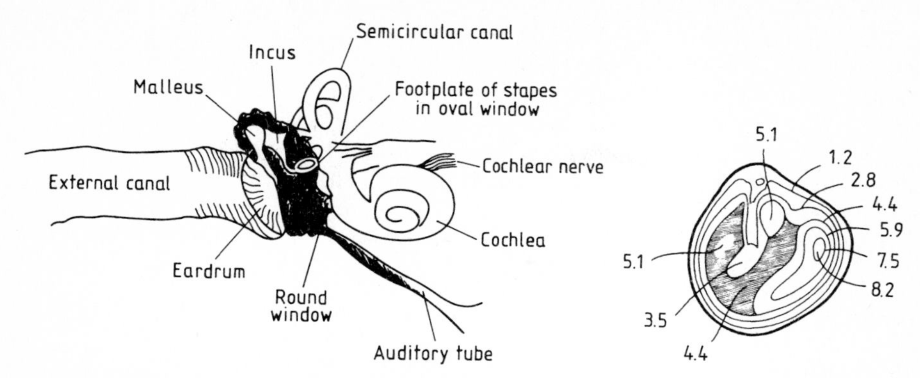

Fig. 2.16. Human Ear. Sound waves impinging on pinna enter external ear canal and travel to eardrum at entrance to middle ear. Middle ear contains three ossicles which act as impedance transformer between airborne sound in external ear and liquid-borne sound in inner ear [2.21]

Fig. 2.17. Contour lines of equal amplitude of human eardrum vibration at 525 Hz (121 dB SPL). Each value must be multiplied by 10^{-5} cm [2.23]

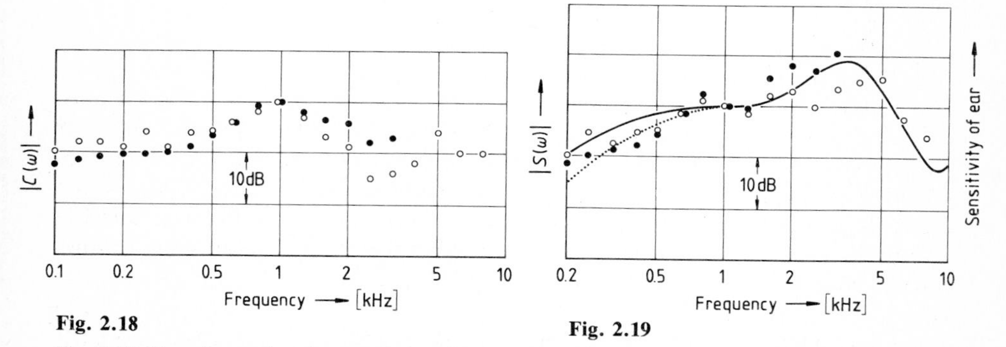

Fig. 2.18

Fig. 2.19

Fig. 2.18. Transformation characteristics of human middle ear between the sound pressure at the eardrum and the apparent pressure on the cochlea. (●) Average values measured [2.17]; (○) measured values [2.24]

Fig. 2.19. Sensitivity of human ear to a sound source located in front of listeners. (———) Normal hearing threshold (after ISO recommendation); (⋯⋯) reexamined low frequency range [2.25]; (●, ○) transformation characteristics between the sound pressure for the sound source and the pressure to the cochlea, $|S(\omega)| = |H(\omega)E(\omega)C(\omega)|$. Data obtained from measured values by *Onchi* (●) [2.17] and *Rubinstein* (○) [2.24] (Fig. 2.18), combined with the transfer function $|H(\omega)E(\omega)|$ measured by *Mehrgardt* and *Mellert* [2.18] (Fig. 2.15)

Khanna [2.23] measured the vibration pattern by time-averaged holography, which allows finer vibration patterns on the eardrum to be perceived, as shown in Fig. 2.17. Note that the outline of the malleus is visible in the pattern at the value of 3.5. The vibration on the malleus is transmitted to the incus and the stapes. The transfer function $C(\omega)$ of the human middle ear between the sound pressure at the eardrum and the apparent sound pressure on the cochlea is plotted in Fig. 2.18. The values are rearranged of the author. Data was measured by *Onchi* [2.17] and *Rubinstein* [2.24] from cadavers. The maxima at 1 kHz are adjusted to the same value. The transfer function of the middle ear varies with sound level due to the acoustic reflex in a living body, which is the combined action of the stapedius and the tympanic muscles. We shall have to ignore such nonlinearity, since data from cadavers only is available.

For weak sounds, the transfer function between a sound source located in front of the listener and the cochlea may be represented by

$$S(\omega) = H(0,0,\omega)E(\omega)C(\omega) . \tag{2.48}$$

The values are plotted in Fig. 2.19. The pattern of the transfer function agrees well with the ear sensitivity for people with normal hearing, so that ear sensitivity may be primarily characterized by the transfer function from the free field to the cochlea [2.26]. It is interesting to compare this with the values reexamined in the low-frequency range [2.25].

2.5.3 Cochlea

The stapes is the last bone of the three auditory ossicles, and the smallest bone of the human body. It is connected with the oval window, and drives the fluid in the cochlea, producing a traveling wave along the basilar membrane. The cochlea contains the sensory receptor organ, i.e., the hair cells on the basilar membrane, which transforms the fluid vibration into the neural code, Fig. 2.20. The basilar membrane is so flexible that each section can move independently of the neighboring section. Thus, each section of basilar membrane can be represented by a single spring-mass system with a point impedance $Z(x, \omega)$ given by

$$Z(x, \omega) = \mathrm{j}\,\omega m(x) + r(x) + s(x)/\mathrm{j}\,\omega , \tag{2.49}$$

where x is the position along the membrane, $m(x)$ specifies the membrane mass, $r(x)$ is the acoustic resistance and $s(x)$ is the stiffness of the membrane. The velocity of the basilar membrane $V(x, \omega)$ is expressed by

$$V(x, \omega) = \frac{-2P(x, \omega)}{Z(x, \omega)} , \tag{2.50}$$

where $2P(x, \omega)$ is the pressure across the membrane.

Equation (2.49) can be rewritten in electrical form [2.28]

$$Z(x, \omega) = \mathrm{j}\,\omega L(x) + R(x) + 1/\mathrm{j}\,\omega C(x) , \tag{2.51}$$

as shown in Fig. 2.21. Near the stapes $0 < x < x_r$, it becomes

$$Z(x, \omega) \approx 1/\mathrm{j}\,\omega C(x) . \tag{2.52}$$

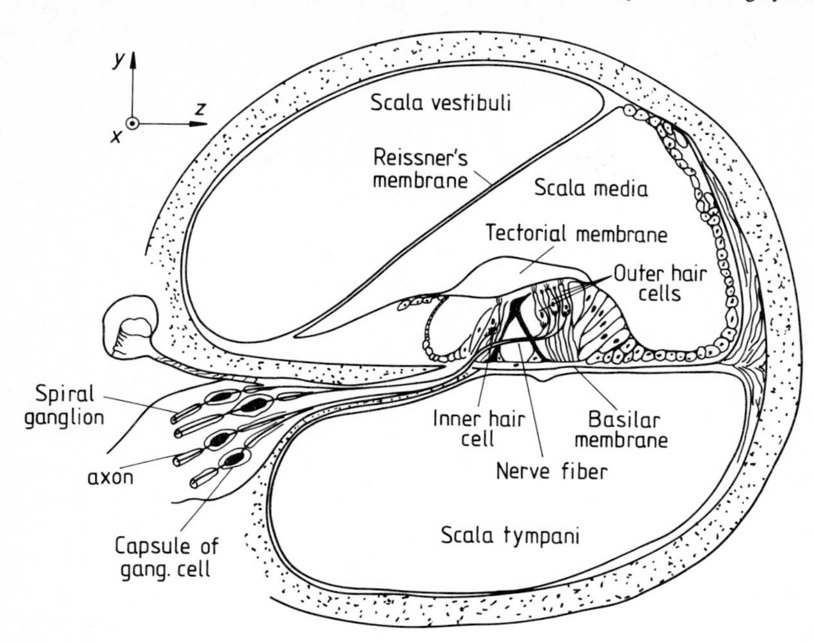

Fig. 2.20. Cross section through cochlea showing fluid-filled canals and basilar membrane supporting hair cells [2.27]

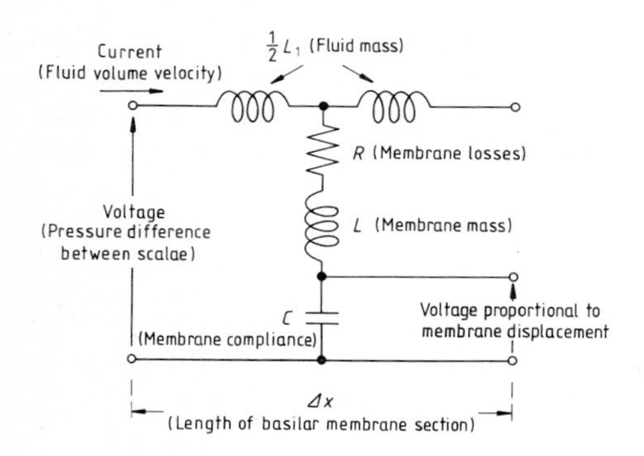

Fig. 2.21. Electric analog circuit for short section of basilar membrane [2.28]

The transmission line behaves essentially as a delay line with a delay of $(L_1(x)C(x))^{1/2}$ per unit section, where $L_1(x)$ is defined in Fig. 2.21. At the characteristic section $x = x_r$,

$$Z(x) = R(x) \tag{2.53}$$

and the velocity of the basilar membrane given by (2.50) becomes maximum.

For $x_r < x < 35$ mm,

$$Z(x, \omega) \approx \mathrm{j}\,\omega L(x)\,. \tag{2.54}$$

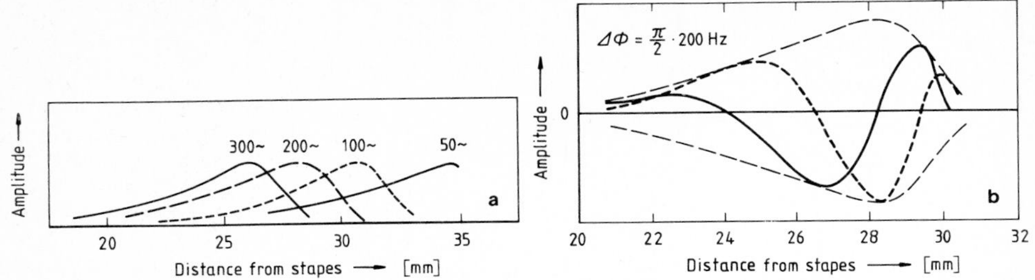

Fig. 2.22. (**a**) Envelopes of traveling waves on basilar membrane [2.22]; (**b**) Traveling waves on basilar membrane for 200 Hz tone [2.22]

Thus, the line behaves as an inductive attenuator and therefore rapid attenuations in this region are observed.

The traveling waves on the basilar membrane observed by *Békésy* [2.22], Fig. 2.22a, b, are consistent with this representation. Further information on cochlear mechanics and modeling may be found in [2.29, 30].

2.6 Nervous System

Physiological signal processing has seldom been discussed in concert-hall acoustics, but the nervous system is one of the most important parts in the whole acoustic system. It is logical that there must be some deep connections between sound signals and subjective attributes.

The mechanical information in the traveling waves on the basilar membrane is transduced into biological information. The transducers, consisting of 15000 receptors on the basilar membrane, are specialized nerve cells called hair cells. As an aid in the understanding of this transducing mechanism, as well as the conduction and transmission of information in the nervous system, a few general mechanisms of the nervous system will be described.

Between the inside and outside of a neural cell (also in the axon) membrane, there is a potential difference of about 70 mV (inside negative, −70 mV) which is called a resting (membrane) potential, as shown in Fig. 2.23. The potential arises mainly from the high concentration of K^+ ions in the cell and the selective permeability of the cell membrane for this ion. The nerve cell can be activated when it is depolarized to the critical level (about −50 mV). An "action potential" is elicited with a short-lasting (about 1 ms) potential charge up to about +30 mV. The action potential is caused by the rapid increase of the permeability of Na^+ ions which are much more concentrated outside the cell than inside. After the activation, an absolute refractory or resting period of about 1 ms exists which determines the maximum firing rate. The high concentration of Na^+ ions on the outside of the cell and K^+ ions inside the cell is maintained by the "Na-K pump" mechanism, in which the energy needed is

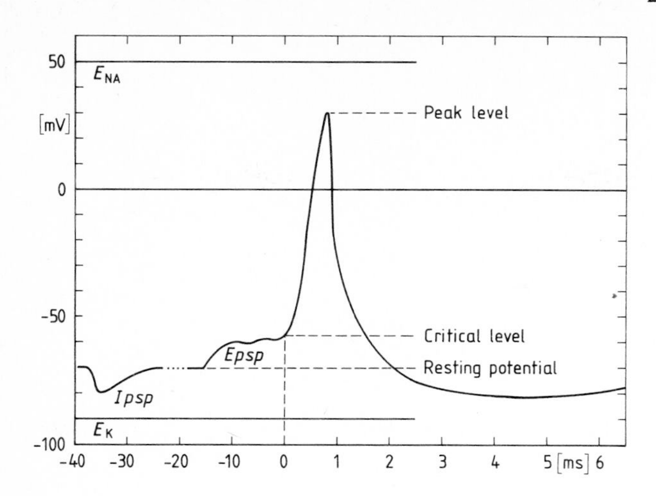

Fig. 2.23. Potential difference between the inside and outside of a neural cell and an action potential. (IPSP: Inhibitory postsynaptic potential; and EPSP: Excitatory postsynaptic potential)

supplied by adenosine triphosphate (ATP). The resting potential is a little less than the equilibrium potential of a K^+ ion, given by

$$E_K = \frac{RT}{F} \ln \frac{[K^+]_o}{[K^+]_i} \approx -90 \text{ mV}, \tag{2.55}$$

where F is the Faraday constant, R is the gas constant and T signifies the absolute temperature. Symbols $[\ ]_o$ and $[\ ]_i$ indicate the concentrations of the ion outside and inside the cell, respectively. The peak value of the action potential approaches the equilibrium potential of the Na^+ ion, given by

$$E_{Na} = \frac{RT}{F} \ln \frac{[Na^+]_o}{[Na^+]_i} \approx 50 \text{ mV}, \tag{2.56}$$

but the resulting peak level is limited at about +30 mV.

The action potential is conducted through the axon (nerve fiber) which is split into many branches (Fig. 2.24a). Without direct connection the axonal endings contact the dendrite and the cell body itself of the next secondary neuron. This contact site with a narrow cleft (ca. a 10 nm gap) is called a synapse. When the action potential impulse reaches the synapse, a chemical substance, called a transmitter, is ejected into the synaptic cleft from the axon terminals, and reaches the subsynaptic site through diffusion, causing a change in the membrane potential of the neuron. The time needed for this transmission, called the synaptic delay, is about 0.5 ms or less. This may be operative as an element of the correlation process. There are two kinds of synapses, excitatory and inhibitory. The former depolarize the membrane potential and cause excitatory postsynaptic potential (EPSP) and the latter hyperpolarize the cell by inhibitory postsynaptic potential (IPSP) (Fig. 2.23). If much EPSP arrives at the cell in a short time, then the potential rapidly reaches the critical level of the activation. Generally speaking, on the cell body

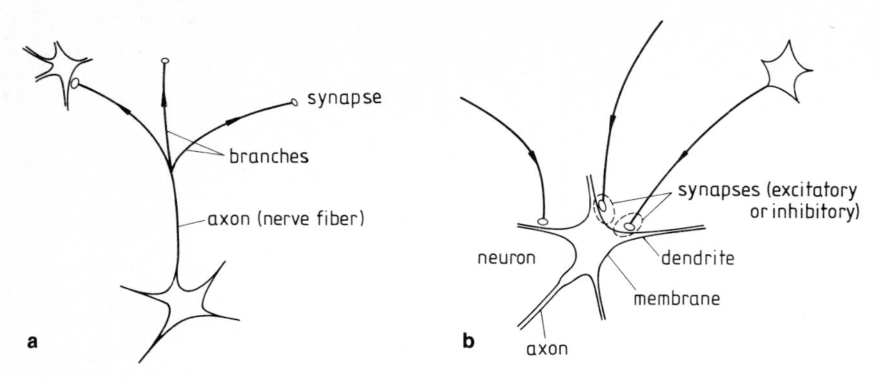

Fig. 2.24a, b. Connections of neurons. **(a)** Divergence; **(b)** convergence

and on the dendrites of a neuron there are a number of excitatory and inhibitory synapses from a number of other neurons. This is called convergence (Fig. 2.24b). When a sum of the EPSP and IPSP reaches the critical level, the secondary neuron fires and the action potential is conducted. Thus, the neural signal results in a certain delay; it is processed at the neuron and it will be transmitted further through the axon branches to a number of third neurons (divergence).

When the hairs of hair cells are bent mechanically, the membrane potential of the hair cell changes. This is called generator of receptor potential. Note that the cochlear microphonics may be a result of accumulated generator potential of many hair cells. There is a logarithmic relation between the strength of the stimulation and the amplitude of the generator potential possibily at least partially related to the Weber-Fechner law [2.31]. Regarding the amplitude of the generator potential, the transmitter is ejected from the hair cell and is transfused to the subsynaptic receptor of the terminal of the secondary cell in the cochlear nuclei through the spiral ganglion, becoming the source of the EPSP. When the EPSP reaches the critical level, the nerve fiber fires. The frequency of the action potential has a linear relation with the amplitude of the EPSP. Thus the frequency of the action potential is roughly a linear function of the logarithm of the stimulus intensity.

The action potentials from the hair cells are conducted and transmitted to a higher level in the above-mentioned manner. The frequency response functions, called "tuning curves" of a single fiber, were first systematically demonstrated by *Katsuki* and his group [2.32] in the auditory pathway. The results of the threshold response in the potential activity of the cochlear nerve of a cat are shown in Fig. 2.25a, and of the trapezoid body in Fig. 2.25b. An important phenomenon is the so-called sharpening effect. The tuning curve becomes sharper than the resonance curve on the basilar membrane. This tendency becomes more distinct at higher levels. *Békésy* [2.33] explained this as a result of an inhibitory action and a funneling effect of neural networks. Interactions

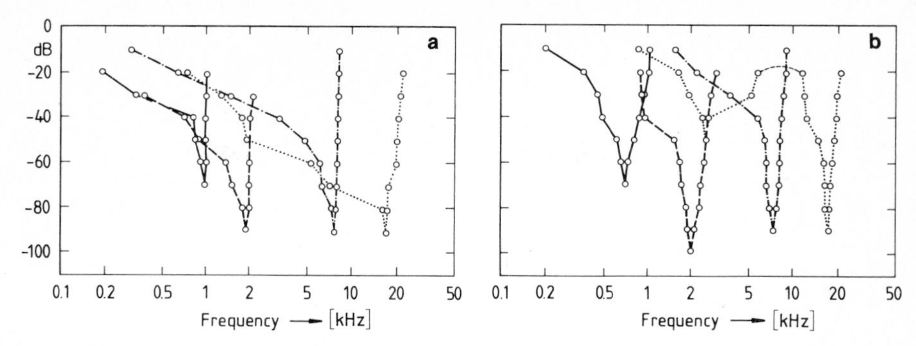

Fig. 2.25 a, b. Frequency response functions of single fibers as threshold responses in the potential activity of a cat's auditory system: Each line with different symbol indicates the response of different single fiber. **(a)** Cochlea nerve; **(b)** trapezoid body [2.32]

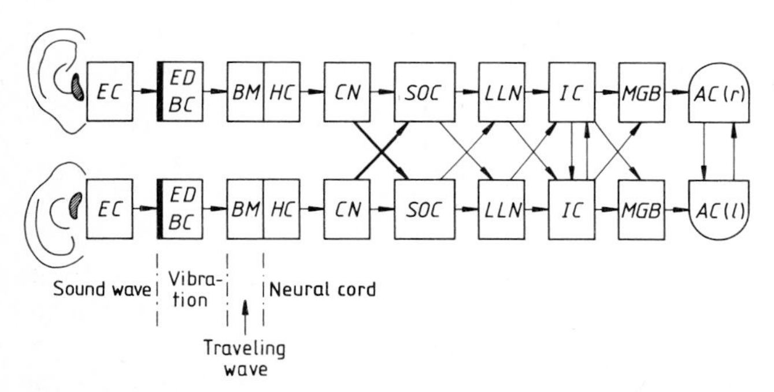

Fig. 2.26. Auditory pathways. (EC: External canal; ED and BC: eardrum and bone chain; BM and HC: basilar membrane and hair cell; CN: cochlea nuclei; SOC: superior olivary complex (medial superior olive, lateral superior olive and nucleus of the trapezoid body); LLN: lateral lemniscus nucleus; IC: inferior colliculus; MGB: medial geniculate body; and AC: auditory cortex of the right and left hemispheres). Feedback mechanisms are not considered

between neighboring neurons are responsible at least partially for the sharpening. Therefore, responses of a single pure tone ω tend to approach a limited region in the auditory pathway x', i.e.,

$$\omega \rightarrow x' \ . \tag{2.57}$$

Accordingly, the input power density spectrum of the cochlea $I(\omega)$ can be roughly mapped at the nerve position x', so that the spectrum can be written as $I(x')$. This neural activity, in turn, may approximately attain the autocorrelation function.

Beyond this point, however, knowledge of the organic functions in the auditory pathway seems to be limited at the present. In addition to the cochlear nuclei, there are the superior olivary complex, the lateral lemniscus nuclei, the inferior colliculus and the medial geniculate body, Fig. 2.26. For further reading in this area, see [2.34]. It is possible that neural signals are processed at every relay station, and therefore, in considering the mechanism of the nervous system, an appropriate model for the auditory pathway between the cochlear nuclei and the auditory cortex should be formed. Since several interaural cross connections are known to exist as physiological structures, in particular, a great number of neurones at and beyond the superior olivary complex are responsive to timing and intensity difference at the two ears, it is therefore quite plausible to assume that they act together as some sort of interaural cross-correlation processor. A model which includes autocorrelation processors is discussed in Sect. 5.1 (Fig. 5.1).

3. Simulation of Sound Fields

By simulations with the aid of a digital computer, we may easily adjust any of the objective parameters of the sound fields and hope to find the optimal conditions for the parameter by subjective judgments. This chapter describes a method of simulating sound fields in a listening room to perform preference judgment.

3.1 Signals at Both Ears

To simplify the analysis, let us first consider a single sound source on a stage. Let $h_l(r|r_0;t)$ and $h_r(r|r_0;t)$ be the pressure impulse responses between the sound source located at r_0 and the left and the right ear-canal entrances, respectively, of a listener, where r is located at the center of the head. Then, the pressure at both ears, which must include all acoustic information, is expressed by

$$f_l(t) = \int_{-\infty}^{t} p(v)\,h_l(t-v)\,dv = p(t) * h_l(t)\,,$$
$$f_r(t) = \int_{-\infty}^{t} p(v)\,h_r(t-v)\,dv = p(t) * h_r(t)\,, \tag{3.1}$$

where $p(t)$ is a source signal and the asterisk denotes convolution. The impulse responses may be decomposed into a set of the impulse responses $w_n(t)$ describing the reflection property of boundaries and the impulse responses from the free field reflections to the ear-canal entrance $h_{nl}(t)$ or $h_{nr}(t)$, n denoting a single sound reflection with a horizontal angle ξ ($\xi = 0$, frontal direction) and an elevation angle η ($\eta = 0$, horizontal plane) relative to the listener, $n = 0$ refers to the direct sound. (The head-related impulse responses $h_{l,r}(t)$ are expressed by the inverse Fourier transform of $H_{l,r}(r|r_0;\omega)$, defined in Sect. 2.5.1.)

The impulse response may now be written as

$$h_{l,r}(r|r_0;t) = \sum_{n=0}^{\infty} A_n w_n(t - \Delta t_n) * h_{nl,r}(t)\,, \tag{3.2}$$

where A_n and Δt_n are the pressure amplitude and the delay time of the reflections relative to the direct sound. The amplitude A_n is determined by the

"$(1/r)$ law", A_0 being unity. Every n ($\geqq 1$) corresponds to a single reflection with azimuth angle ξ and elevation angle η to the listener. Equation (3.1) becomes

$$f_{l,r}(t) = \sum_{n=0}^{\infty} p(t) * A_n w_n(t - \Delta t_n) * h_{nl,r}(t) \,. \tag{3.3}$$

If the sound source has non-uniform radiation, the radiation pattern of the sound source may be broken down into a number of discrete beams. For example, $p(t)$ may be replaced by $p_n(t)$ in the above equation. If there are many sound sources distributed on the stage, then the pressures at the ears may be expressed as a linear sum of $f_{l,r}(t)$ given by (3.3) for usual sound pressure levels. (The preferred condition for the distributed sources may be found by aggregating the preferred conditions for the individual and important sources on the stage.)

All independent objective parameters of acoustic information, which must be contained in the sound pressures present at the two ears, as given by (3.3), may be reduced to the following [3.1].

i) The first parameter is the source signal $p(t)$, which can be represented by its long-time autocorrelation function as defined by (2.9),

$$\Phi_p(\tau) = \lim_{T \to \infty} \frac{1}{2T} \int_{-T}^{+T} p'(t) p'(t+\tau) dt \,,$$

where $p'(t) = p(t) * s(t)$. The ear sensitivity $s(t)$ may be characterized by the external ear and the middle ear as in Sect. 2.5.2 as $[s(t) \simeq h_{0l,r}(t) * e(t) * c(t)]$. For practical convenience, $s(t)$ is chosen as the impulse response of an A-weighting filter. The autocorrelation function may be factored into the intensity of the sound signal $\Phi_p(0)$ and the normalized autocorrelation function, defined by

$$\phi_p(\tau) = \frac{\Phi_p(\tau)}{\Phi_p(0)} \,. \tag{3.4}$$

For several music motifs and a speech signal, measured autocorrelation functions are shown in Fig. 2.2 of Sect. 2.1 (*temporal-monaural criterion*).

ii) The second objective parameter is the set of impulse responses of the reflecting boundaries, $A_n w_n(t - \Delta t_n)$. These impulse responses represent the initial time-delay gap between the direct sound and the first reflection as well as the structure of the early reflections, the subsequent reverberation, and any spectral changes due to the reflections. Note that factors A_n and Δt_n are expressed in terms of distance to the sound source; thus these two factors are not independent of each other (*temporal-monaural criterion*).

iii) The two sets of the head-related impulse responses for the two ears $h_{nl,r}(t)$ constitute the remaining objective parameter. These responses $h_{nl}(t)$ and $h_{nr}(t)$ play an important role in localization, but are not mutually independent objective factors. For example, $h_{nl}(t) \approx h_{nr}(t)$ in the median plane ($\xi = 0°$).

Therefore, to represent the interdependence between these impulse responses, one may introduce a single factor, i.e., the long-time interaural cross correlation between the continuous sound signals $f_l'(t)$ and $f_r'(t)$. This becomes a significant factor in determining the degree of subjective diffuseness of sound fields [3.2–4] (Appendix A). Subjective diffuseness (the lack of a special directional impression) is perceived when listening to sound fields with a low degree of interaural cross correlation. On the other hand, a well-defined direction is perceived if the interaural cross correlation has a strong peak for $|\tau| < 1$ ms. The interaural cross correlation depends mainly on the directions from which reflections arrive at the listener and on their amplitudes (*spatial-binaural criterion*).

The interaural cross correlation is defined by

$$\Phi_{lr}(\tau) = \lim_{T\to\infty} \frac{1}{2T} \int_{-T}^{+T} f_l'(t) f_r'(t+\tau)\,dt\,, \qquad |\tau| \leqq 1 \text{ ms}\,. \tag{3.5}$$

First, let us consider the interaural cross correlation $\Phi_{lr}^{(0)}(\tau)$ of the direct sound only. The pressure at each ear (cochlear) is then expressed by

$$f_l'(t) = p'(t) * h_{0l}(t)\,,$$
$$f_r'(t) = p'(t) * h_{0r}(t)\,.$$

If the listener is facing the sound source, the normalized interaural cross correlation defined by

$$\phi_{lr}^{(0)}(\tau) = \frac{\Phi_{lr}^{(0)}(\tau)}{\sqrt{\Phi_{ll}^{(0)}(0)\,\Phi_{rr}^{(0)}(0)}} \tag{3.6}$$

approaches unity because of $h_{0l}(t) \approx h_{0r}(t)$, where $\Phi_{ll}^{(0)}(0)$ and $\Phi_{rr}^{(0)}(0)$ are the autocorrelation functions at $\tau = 0$ for each ear.

If discrete reflections are added to the direct sound after the autocorrelation function of the direct sound becomes weak enough, the normalized interaural cross correlation is expressed by

$$\phi_{lr}^{(N)}(\tau) = \frac{\sum_{n=0}^{N} A_n^2 \Phi_{lr}^{(n)}(\tau)}{\sqrt{\sum_{n=0}^{N} A_n^2 \Phi_{ll}^{(n)}(0) \sum_{n=0}^{N} A_n^2 \Phi_{rr}^{(n)}(0)}}\,, \qquad w_n(t) = \delta(t)\,, \tag{3.7}$$

where $\Phi_{lr}^{(n)}(\tau)$ is the interaural cross correlation of the nth reflection, $\Phi_{ll}^{(n)}(0)$ and $\Phi_{rr}^{(n)}(0)$ are autocorrelation functions at $\tau = 0$ of the nth reflection at the ears, and $\delta(t)$ is the Dirac delta function. If $w_n(t) \neq \delta(t)$, then A_n may be approximately rewritten as

$$A_n \int_{\Omega_1}^{\Omega_2} |P'(\omega)||W_n(\omega)|d\omega / \int_{\Omega_1}^{\Omega_2} |P'(\omega)|d\omega ,$$

where Ω_1 and Ω_2 are the lower and upper limits of the audible frequencies (rad/s), and $P'(\omega)$ and $W_n(\omega)$ are the Fourier transforms of $p'(t)$ and $w_n(t)$, respectively.

These correlations at the two ears have not yet been theoretically obtained. Therefore, the long-time interaural cross correlations ($2T = 35$ s) were measured for each single reflected sound direction arriving at a dummy head "listener". The dummy head was constructed according to acoustic measurements of the auditory threshold level, so that the output signals of the microphones corresponded to the ear sensitivity [3.5].

Measured values of the interaural cross correlation using music motifs A and B (Table 2.1) are shown in Fig. 3.1 for different horizontal angles to the

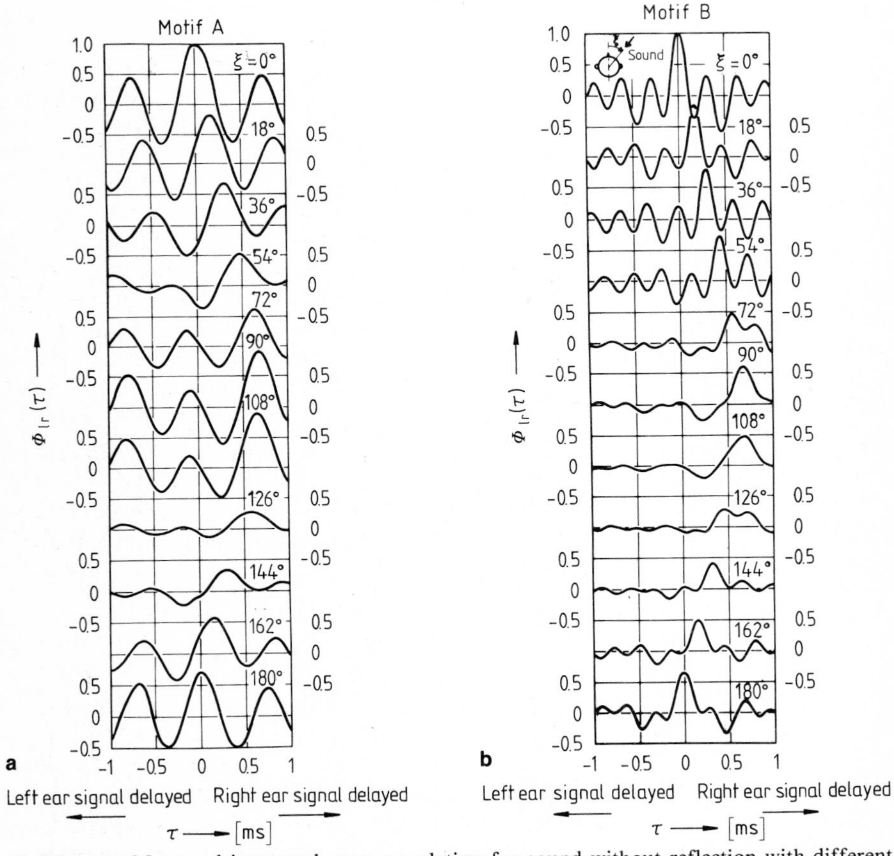

Fig. 3.1a, b. Measured interaural cross correlation for sound without reflection with different horizontal angles of incidence ξ [3.1]. **(a)** Music motif A, **(b)** music motif B

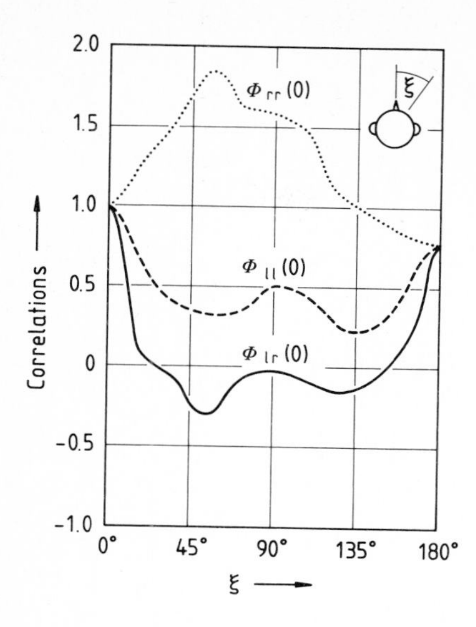

Fig. 3.2. Average values of the interaural cross correlation and the autocorrelation functions at each ear ($\tau = 0$) arriving from ξ for single music signal, $\Phi_{rr}(0)$ and $\Phi_{ll}(0)$ correspond to the sound energies at the right and left ears, respectively

sound source. The geometrical arrangement is shown in the upper part of Fig. 3.1 b. The values of the interaural cross correlation for $|\tau| \leqq 1$ ms and autocorrelation functions at $\tau = 0$ are listed in Table D.1 (Appendix D), as well as those of other sound signals. For convenience, autocorrelation functions at $\tau = 0$ for the frontal incidence $\xi = 0°$ are adjusted to unity in Fig. 3.1 and Table D.1. These values are used to calculate the interaural cross correlation by (3.7). The values of $\Phi_{ll}(0)$ and $\Phi_{rr}(0)$ correspond to the average intensities of the sound at the left and right ears, respectively. The interaural cross correlation $\Phi_{lr}(\tau)$ is maximum at $\tau = 0$ for typical sound fields of concert halls. The average values at $\tau = 0$ for five music signals are shown in Fig. 3.2 and Table D.2. Similar behavior of interaural cross correlation is commonly found for all the music signals; for example, values of $\Phi_{lr}(0)$ show maxima at the frontal incidence and minima at $\xi \approx 55°$. Furthermore, large differences are obtained between the values of $\Phi_{ll}(0)$ and $\Phi_{rr}(0)$ at $\xi = 55°$ because of the amplitude differences between the ears. Reflections arriving in the median plane ($\xi = 0°$, $\xi = 180°$) all result in large IACC values (≈ 1.0).

The magnitude of the interaural cross correlation is defined by

$$\mathrm{IACC} = |\phi_{lr}(\tau)|_{\max}, \quad \text{for} \quad |\tau| \leqq 1\ \mathrm{ms}\,. \tag{3.8}$$

The calculated IACCs of sound fields with a single reflection are shown in Fig. 3.3 as a function of the horizontal direction of the reflection. A similar result is obtained when values at $\tau = 0$ are plotted as shown in Fig. 3.4. The behavior of the IACC as a function of the angle of arrival is similar for all the music motifs as well as for the noise source. Thus the IACC magnitude is almost independent of the source signal. Inspection of Figs. 3.3, 4 indicates that the IACC drops rapidly with increasing ξ to a minimum for discrete

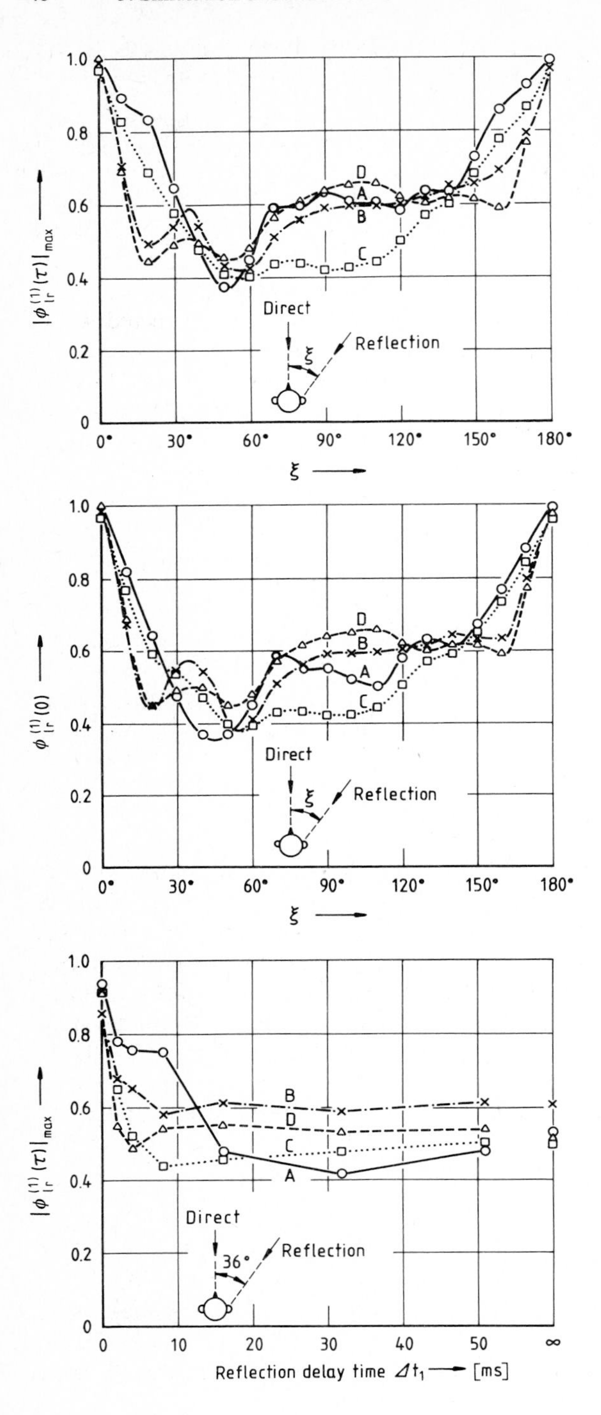

Fig. 3.3. Calculated maximum absolute values of the interaural cross correlation (IACC) using (3.8) of music sound fields with single reflection [3.1]. A – D: different music motifs

Fig. 3.4. Calculated values of interaural cross correlation at $\tau = 0$ of music sound fields with single reflection. Similar results as those indicated in Fig. 3.3 are found [3.1]. A – D: different music motifs

Fig. 3.5. Measured values of the IACC of the music sound fields as a function of delay of the single reflection ($\xi = 36°$) [3.1]. Values at $\Delta t_1 \to \infty$ are calculated by (3.8). A – D: different music motifs

reflections in the range $\xi = 15°$ to $60°$ and levels off for higher angles. It will be shown in Sect. 4.2 that this angular range is also of special importance subjectively.

To determine the minimum delay time required for (3.7) to hold, measured maximum values of the interaural cross correlation as a function of delay time Δt_1 are plotted in Fig. 3.5. Calculated values are shown for $\Delta t_1 \rightarrow \infty$, which agree well with the measured values at all delay times except for $\Delta t_1 < 4$ ms (motifs B, C, and D) and $\Delta t_1 \leqq 15$ ms (motif A). When the relative delay of reflections $\Delta t_{n+1} - \Delta t_n$ approaches zero, the measured values of IACC generally increase. A subjective impression of diffuseness may be expected when the reflection arrives more than 4 ms later. This condition usually occurs in early reflections in concert halls, so the IACC may be calculated prior to construction of a new hall. It is shown in Sect. 7.3 that the calculated values of IACC converge rapidly to the final value when only the first few early reflections are taken into account and that these agree with measured values.

If there is a sound field with the direct sound ($A_0 = 1.0$, $\xi = 0°$, $\eta = 0°$) and two reflections ($A_1 = 0.8$, $\xi = 54°$, $\eta = 0°$; $A_2 = 0.6$, $\xi = 342°$, $\eta = 0°$) of the music motif B, then the IACC ($\tau = 0$) is approximately calculated by

$$\phi_{lr}^{(2)}(0) \approx \frac{0.99 - 0.18 - 0.06}{\sqrt{(1.0 + 0.22 + 0.50)(1.0 + 1.32 + 0.19)}} \approx 0.36 \, .$$

In this calculation, the suffices l and r are interchanged for the second reflection with $\xi = 18°$, and values of correlations listed in Table D.1b are used.

3.2 Simulation of Sound Localization

As mentioned above, the directional information of the sound must be taken into consideration in simulating sound fields in concert halls. For this purpose, *Schroeder* and *Atal* [3.6] first simulated sound localization in the horizontal plane by a two-loudspeaker reproduction system. To make the perception correspond more precisely to the actual direction of a sound source located at any position in a three-dimensional space, a general system considering asymmetry of our head and pinnae is described as follows [3.7].

Referring to the lower part of Fig. 3.6, let the pressure impulse responses for the paths from the two loudspeakers L_1 and L_2 to the entrances of the left and right ear canals be $h_{l,r1}(t)$ and $h_{l,r2}(t)$, respectively. Then, the pressures to be reproduced at the two ears are expressed by

$$\begin{aligned} f_l(t) &= x_1(t) * h_{l1}(t) + x_2(t) * h_{l2}(t) \, , \\ f_r(t) &= x_1(t) * h_{r1}(t) + x_2(t) * h_{r2}(t) \, , \end{aligned} \tag{3.9}$$

where $x_1(t)$ and $x_2(t)$ are the input signals supplied for the loudspeakers L_1 and L_2, respectively.

Fourier transforming both sides of the above equation gives

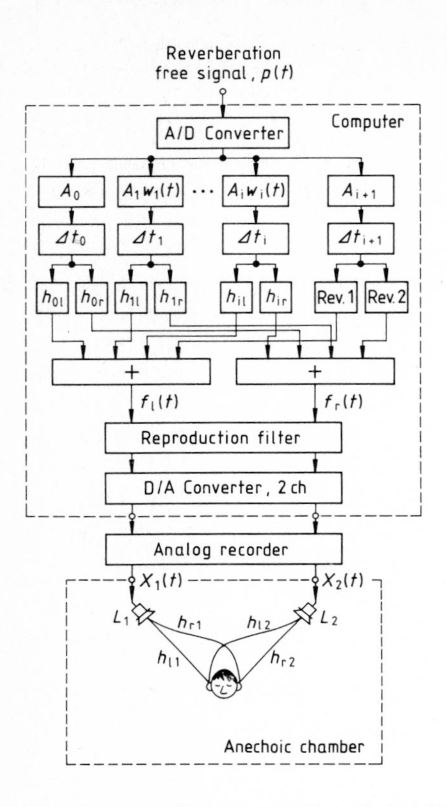

Fig. 3.6. Block diagram of system for simulating sound fields in concert halls with two-loudspeakers

$$
\begin{aligned}
F_l(\omega) &= X_1(\omega)H_{l1}(\omega) + X_2(\omega)H_{l2}(\omega)\,, \\
F_r(\omega) &= X_1(\omega)H_{r1}(\omega) + X_2(\omega)H_{r2}(\omega)\,.
\end{aligned}
\tag{3.10}
$$

The input signals to the loudspeakers in the frequency domain are derived by

$$
\begin{bmatrix} X_1(\omega) \\ X_2(\omega) \end{bmatrix} = D(\omega)^{-1} \begin{bmatrix} H_{r2}(\omega) & -H_{l2}(\omega) \\ -H_{r1}(\omega) & H_{l1}(\omega) \end{bmatrix} \begin{bmatrix} F_l(\omega) \\ F_r(\omega) \end{bmatrix}, \tag{3.11}
$$

where

$$
D(\omega) = H_{l1}(\omega)H_{r2}(\omega) - H_{r1}(\omega)H_{l2}(\omega)\,. \tag{3.12}
$$

Therefore, the signals in the time domain are obtained by the inverse Fourier transform, so that

$$
\begin{aligned}
x_1(t) &= [f_l(t) * h_{r2}(t) - f_r(t) * h_{l2}(t)] * d(t)\,, \\
x_2(t) &= [f_r(t) * h_{l1}(t) - f_l(t) * h_{r1}(t)] * d(t)\,,
\end{aligned}
\tag{3.13}
$$

where $d(t)$ is the inverse Fourier transform of $D(\omega)^{-1}$. The necessary and sufficient condition for a unique solution is $D(\omega) \neq 0$, for the full frequency range.

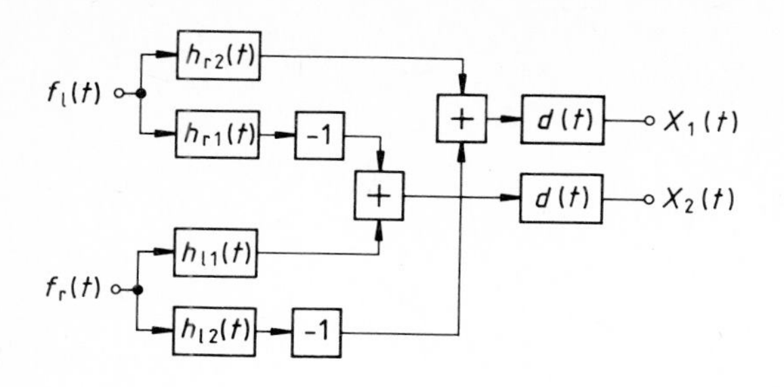

Fig. 3.7. Block diagram of the reproduction filter for two-channel loudspeaker systems (asymmetry): $f_l(t)$ and $f_r(t)$ are signals to be reproduced at both ears; $x_1(t)$ and $x_2(t)$ are signals to be supplied to two loudspeakers

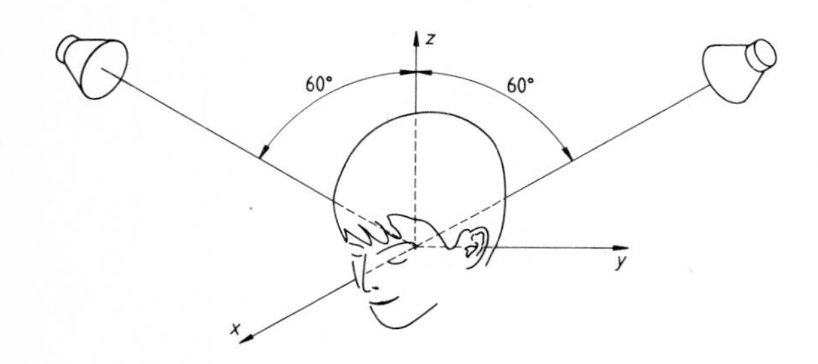

Fig. 3.8. Optimum loudspeaker system for simulating sound localization in three-dimensional space

From (3.13), it is easy to draw a block diagram of the reproduction filter to simulate the sound localization, as shown in Fig. 3.7.

Let us consider a single sound source located at an arbitrary point in a free field. The sound pressure at each ear is

$$f_l(t) = p(t) * h_{nl}(t) , \qquad f_r(t) = p(t) * h_{nr}(t) , \tag{3.14}$$

where $h_{nl,r}(t) \equiv h_{l,r}(\xi, \eta; t)$ are impulse responses between the source and the ear entrances. The signals are reproduced through the filter. The head-related transfer functions required for this purpose are obtained using the single-pulse method (Sect. 7.1). The two reproduction loudspeakers are located above the listener at angles $\pm 60°$ measured from the median plane in the $x-z$ plane (Fig. 3.8) in which the head-related transfer functions are fairly flat with no zeros and no significant dips for all subjects, satisfying the condition of uniqueness. Natural sound localizations with external virtual sound images are created with a minimum resolution of 15°, in both the horizontal and median planes. Responses are shown in Fig. 3.9 as well as localizations with real sound sources [3.8]. In the experiments, a wide-band white noise (300 – 13 600 Hz) is presented as a source signal. By individual transfer functions in the simulation, the accuracy of localization is almost of the same order as for the real sound sources, as shown in the figure.

If we apply the head-related transfer functions from another person, then the subject's accuracy is generally decreased and, in some cases, localization is not possible.

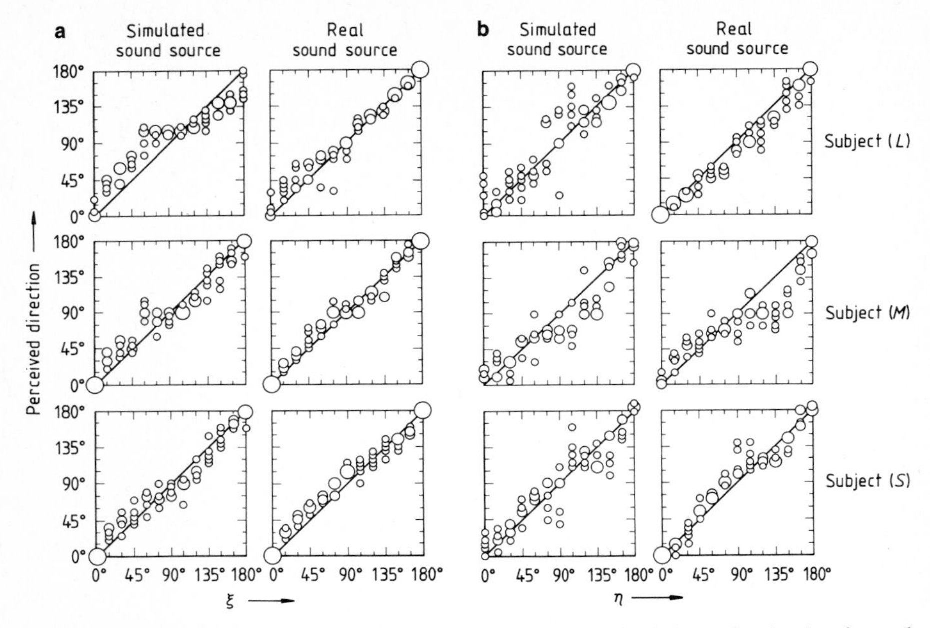

Fig. 3.9a, b. Sound localization by three subjects with different sized pinnae for simulated sound source and for real sound source [3.8]. **(a)** On the horizontal plane; **(b)** on the median plane

3.3 Simulation of Sound Fields in Concert Halls

An example of a simulation system for sound fields in concert halls is illustrated in Fig. 3.6, designed to produce sound pressures according to (3.3). A reverberation-free source signal is fed into a computer through an analog-to-digital converter. The computer program provides the amplitude and the delay of early reflections, including the directional information (reflections from $n = 1, 2, \ldots, i$) and the subsequent reverberation (reflections from $n = i+1, i+2, \ldots$), all calculated relative to the direct sound ($n = 0$). As shown in the first column of the upper part in Fig. 3.6, the direct sound is simulated using only the head-related impulse response to the two ears for the frontal direction, i.e., $p(t) * h_{0l,r}(t)$ with $A_0 = 1$ and $\Delta t_0 = 0$. The second column simulates the first reflection ($n = 1$) for the two ears, which is given by

$$p(t) * A_1 w_1(t - \Delta t_1) * h_{1l,r}(t) \,.$$

Similarly, two early reflections are simulated which can usually be distinguished in the impulse response measured in rooms. After the early reflections, the two incoherent reverberation signals are generated by independent digital reverberators. A block diagram of a reverberator is shown in Fig. 3.10 [3.9].

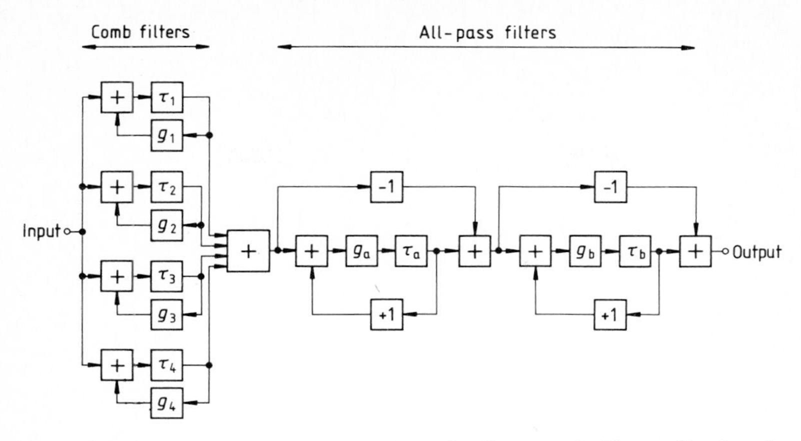

Fig. 3.10. Block diagram of reverberator with four comb filters adjusting the subsequent reverberation time and two all-pass filters producing a proper density of reflections [3.6]

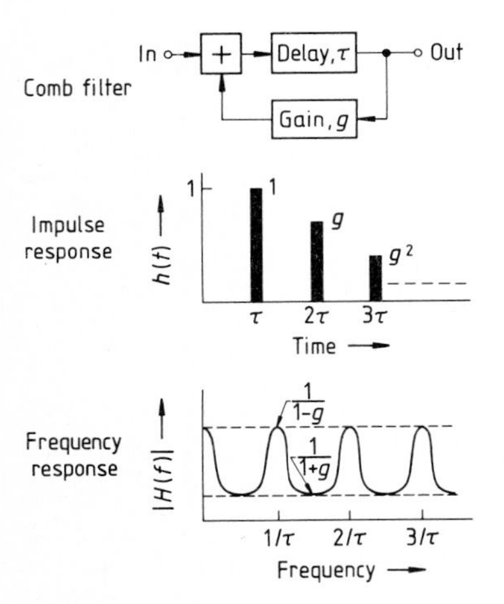

Fig. 3.11. Comb filter, its impulse response and spectrum [3.9]

These sound signals, which are simulated for the left and right ears, are added and sent to the reproduction filter as described in the previous section.

The reverberator consists of comb filters and all-pass filters. The impulse response of one of the comb filters with delay τ and gain g (Fig. 3.11) is expressed by

$$h(t) = \delta(t-\tau) + g\delta(t-2\tau) + g^2\delta(t-3\tau) + \dots \,. \tag{3.15}$$

Thus, the reflections decrease exponentially.
The Fourier transform of (3.15) gives the corresponding signal spectrum:

$$H(\omega) = \mathrm{e}^{-\mathrm{j}\omega\tau} + g\,\mathrm{e}^{-\mathrm{j}2\omega\tau} + g^2\mathrm{e}^{-\mathrm{j}3\omega\tau} + \dots = \frac{\mathrm{e}^{-\mathrm{j}\omega\tau}}{1 - g\,\mathrm{e}^{-\mathrm{j}\omega\tau}} \,. \tag{3.16}$$

The absolute value of $H(\omega)$, which is given by

$$|H(\omega)| = \frac{1}{\sqrt{1+g^2-2g\cos\omega\tau}}, \tag{3.17}$$

is shown in the lower part of Fig. 3.11. The amplitude as a function of frequency presents a comb with periodic structure.

For $\omega = 2n\pi/\tau (n = 0, 1, 2, \ldots)$ and $g > 0$, it has the maxima

$$|H(\omega)|_{\max} = \frac{1}{1-g}, \tag{3.18}$$

and for $\omega = (2n+1)\pi/\tau$, the minima

$$|H(\omega)|_{\min} = \frac{1}{1+g}. \tag{3.19}$$

Therefore, the ratio between the maxima and minima is expressed by

$$\frac{|H(\omega)|_{\max}}{|H(\omega)|_{\min}} = \frac{1+g}{1-g}. \tag{3.20}$$

For example, if $g = 0.85$, then the ratio is $1.85/0.15 = 12.3$ or 22 dB. This produces an undesired "colored" quality. Therefore, the loop gain of the comb filter should not exceed about 0.85. By use of several comb filters connected in parallel, as shown in Fig. 3.10, a highly irregular frequency response is produced similar to that in concert halls. The reverberation time of the reverberator is given by the loop gains $g_1\ g_2, \ldots, g_M$ and delays $\tau_1, \tau_2, \ldots, \tau_M$ of the different comb filters.

A sound level decay by $-20\log g_m$ dB for every trip around the feedback loop τ_m gives

$$T_m = \frac{60}{-20\log|g_m|}\tau_m = \frac{3}{-\log|g_m|}\tau_m, \quad m = 1, 2, \ldots, M, \quad \text{and}$$

$$T_{\text{sub}} = [T_m]_{\max} \tag{3.21}$$

where T_{sub} is the subsequent reverberation time. Note that the reverberation time as a function of frequency can be derived from the impulse response $g_m(t)$ or its frequency domain response $G_m(\omega)$, which corresponds to the transfer function for reflection from boundary walls.

To produce a high density of reflections without changing the frequency response of the reverberator, two all-pass filters are connected in series with the comb filters. The density of reflections at any time t after the impulse excitation is given by

$$n_e(t) = \frac{1}{2}\sum_{m=1}^{M}\frac{1}{\tau_m}\frac{1}{\tau_a\tau_b}t^2. \tag{3.22}$$

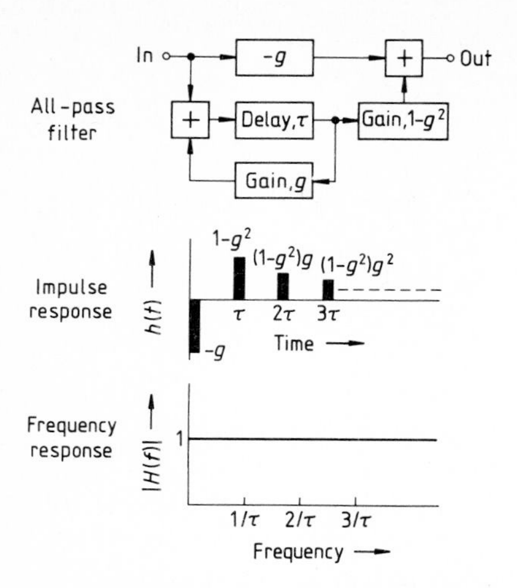

Fig. 3.12. All-pass filter, its impulse response and spectrum [3.9]

The delays τ_a and τ_b of the all-pass filters should be chosen as $\tau_a, \tau_b \ll \tau_m$ ($m = 1, 2, \ldots, M$), so that they do not influence the reverberation time from the comb filters.

To aid analyzing the behavior of the all-pass filter, Fig. 3.12 represents a generalized diagram of the filter. The impulse response is given by

$$h(t) = -g\delta(t) + (1-g^2)[\delta(t-\tau) + g\delta(t-2\tau) + \ldots] . \tag{3.23}$$

Taking (3.16) into consideration, the Fourier transform of (3.23) is given by

$$H(\omega) = -g + (1-g^2)\frac{e^{-j\omega\tau}}{1-g\,e^{-j\omega\tau}} = e^{-j\omega\tau}\left(\frac{1-g\,e^{j\omega\tau}}{1-g\,e^{-j\omega\tau}}\right) . \tag{3.24}$$

Thus,

$$|H(\omega)| = 1.0$$

for all frequencies as shown in the lower part of Fig. 3.12.

In the reverberator shown in Fig. 3.10, if we set $g_a = g_b = 1/\sqrt{2}$ (≈ 0.7), then the all-pass filter is realized.

4. Subjective Preference Judgments

First of all a method is described for obtaining the linear scale value between sound fields by applying of the law of comparative judgment. Paired comparison tests of subjective preference were conducted in relation to the fully independent objective parameters, so that the optimum conditions could be found. If we can obtain a linear scale value of preference for each objective parameter which affects independently on the preference also, then we can calculate a total preference by applying the "principle of superposition".

4.1 Linear Scale Value of Preference

There are several methods of measuring psychological attributes for testing sound fields. As far as the subjective preference is conerned, the paired comparison method is extremely useful, because subjects can simply judge which of two sound fields they prefer to hear. Note that this kind of test avoids all manner of ill-defined epithets such as "clear", "brilliant", "warm", "balanced", "spacious", "intimate", and so on, for which the results have not always been satisfactory, since a change to the acoustics of a room will cause listeners to say that one likeable quality improved while another deteriorated. This new approach to the problem of describing the important qualities preferred by listeners will simplify responses and their interpretation.

After such judgments, preference scores are directly obtained for each pair of sound fields by giving scores of $+1$ and -1, corresponding to positive and negative judgments, respectively. The normalized preference score is obtained by accumulating the scores for all sound fields (F) tested and all subjects (S), and then dividing by the factor $S(F-1)$ for statistical analysis. This kind of data arrangement is convenient for finding a preferred condition of sound fields. This procedure was adopted only for the data from earlier experiments for sound fields with a single reflection (Sects. 4.2.1 – 3).

However, the preference score itself cannot be represented as a linear scale. To obtain the optimum condition, scale value of preference, which is regarded as a linear psychological distance between the sound fields, is derived from *Thurstone's* theory [4.1].

Discriminative processes of preference exist for any two sound fields A and B extracted from F sound fields ($A, B \in F$). These processes can be denot-

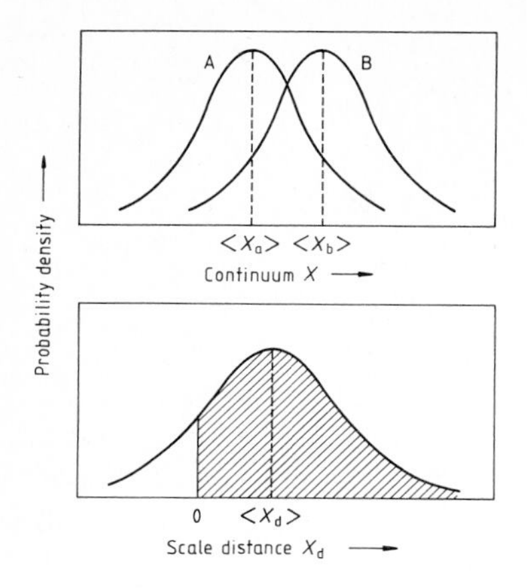

Fig. 4.1. Discriminative processes of subjective preference judgments for two sound fields A, B and a scale distance between the sound fields

ed respectively by X_a and X_b on any particular occasion. Then, the psychological scale is obtained through the frequencies of the discriminative process X, which form a normal distribution on the one-dimensional scale, in which a probability density is given by

$$\frac{1}{\sigma\sqrt{2\pi}} \exp\left(-\frac{(X-\langle X\rangle)^2}{2\sigma^2}\right), \tag{4.1}$$

where σ is the standard deviation. To judge whether A is preferred to B ($A>B$) or B is preferred to A ($B>A$), we introduce a discriminative process $X_\mathrm{d}=X_b-X_a$, as shown in Fig. 4.1 [4.2]. The probability density of X_d also has a normal distribution, so that

$$\frac{1}{\sigma_\mathrm{d}\sqrt{2\pi}} \exp\left(-\frac{(X_\mathrm{d}-\langle X_\mathrm{d}\rangle)^2}{2\sigma_\mathrm{d}^2}\right), \quad \text{where} \tag{4.2}$$

$$\sigma_\mathrm{d}=\sqrt{\sigma_a^2+\sigma_b^2-2\varrho\sigma_a\sigma_b}\,. \tag{4.3}$$

Here ϱ is the correlation coefficient between the discriminative fluctuations in X_a and X_b.

Therefore, the probability that B is preferred to A is expressed by

$$P(B>A)=\frac{1}{\sigma_\mathrm{d}\sqrt{2\pi}}\int_0^\infty \exp\left(-\frac{(X_\mathrm{d}-\langle X_\mathrm{d}\rangle)^2}{2\sigma_\mathrm{d}^2}\right) dX_\mathrm{d}$$

$$=\frac{1}{\sqrt{2\pi}}\int_{Z_{ab}}^\infty \exp\left(-\frac{y^2}{2}\right) dy\,, \quad \text{where} \tag{4.4}$$

$$Z_{ab}=-\langle X_\mathrm{d}\rangle/\sigma_\mathrm{d}\,. \tag{4.5}$$

Using a table of the normal distribution, the value Z_{ab} is found through the probability ($B>A$) from the result of the paired comparison tests. Since $\langle X_d \rangle = \langle X_b \rangle - \langle X_a \rangle$, the scale value is obtained by

$$\begin{aligned}\langle X_a \rangle - \langle X_b \rangle &= Z_{ab}\sigma_d \\ &= Z_{ab}\sqrt{\sigma_a^2 + \sigma_b^2 - 2\varrho\sigma_a\sigma_b}\,. \end{aligned} \tag{4.6}$$

For practical applications, there are five cases, one of which is selected according to the degree of assumptions required to obtain the scale value [4.1]. Throughout the investigations, the magnitude σ_d is used as a unit for the scale of preference. For example, if the probability ($B>A$) is 0.84, then the scale value $\langle X_d \rangle$ is unity.

We similarly obtain the scale values Z_{ab} for all pairs, $a, b = 1, 2, \ldots, F$. The scale value of each sound field is obtained by the method of least squares, minimizing the error, so that

$$\langle X_a \rangle = \frac{1}{F}\sum_{b=1}^{F} Z_{ab}\,, \tag{4.7}$$

where we adjusted the origin of the scale such that

$$\sum_{b=1}^{F} \langle X_b \rangle = 0\,.$$

For further procedures to obtain the scale value, see [4.3 – 5].

4.2 Sound Fields with Single and Multiple Early Reflections

4.2.1 Preferred Delay Time of Single Reflection

Sound with a single reflection of $\xi_1 = 36°$ ($\eta_1 = 9°$) from a fixed direction was selected since this value is a typical reflection angle. In the following section, it is noted that the preferred angle of reflection is somewhat larger, but the relation between preferred time delay and preferred angle is not critical, as clarified below. The delay time was adjusted in the range of 6 – 256 ms. Paired comparison tests were performed for all pairs in an anechoic chamber using healthy subjects. The normalized scores of the sound fields are shown in Fig. 4.2 as a function of the delay.

The most preferred delay time showing the maximum score differs greatly between motifs. If the amplitude of reflection $A_1 = 1$, the most preferred delays are around 130 and 35 ms for motifs A and B, respectively. These correspond to the effective duration of the autocorrelation function as listed in Table 2.1. (For individual differences in the response, see Appendix B).

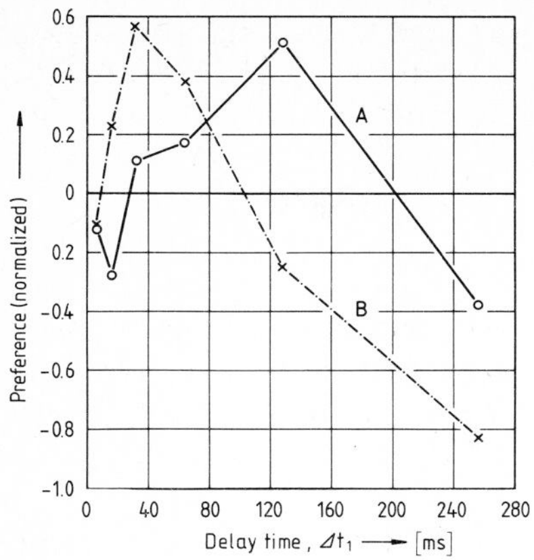

Fig. 4.2

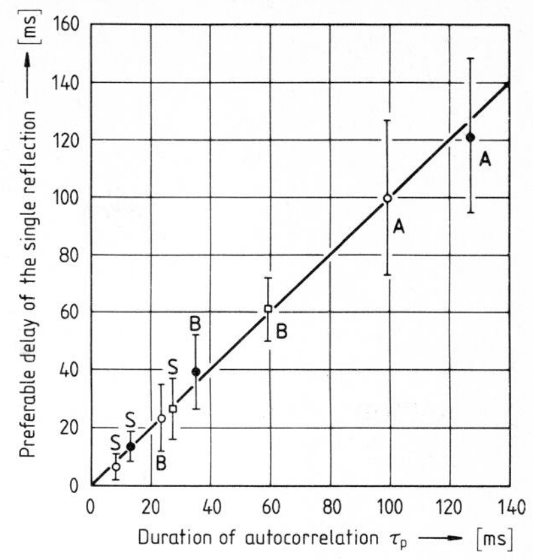

Fig. 4.3

Fig. 4.2. Preference scores of the sound fields as a function of the delay, $\xi = 36°$, $\eta = 9°$, $A_1 =$ 0 dB (6 sound fields and 13 German subjects) [3.1]. A: Music motif A, Royal Pavane by Gibbons, $\tau_e = 127$ ms; B: Music motif B, Sinfonietta, Opus 48, III movement by Malcolm Arnold, $\tau_e =$ 35 ms

Fig. 4.3. Relationship between the preferred delay of the single reflection and the duration of autocorrelation function such that $|\phi_p(\tau)| = 0.1\,A_1$ [3.1]. Ranges of the preferred delay are graphically obtained at 0.1 below the maximum score. A, B and S refer to motif A, motif B and speech, respectively (Table 2.1). Different symbols indicate the center values obtained at the reflection amplitudes of +6 dB (○), 0 dB (●) and −6 dB (□), respectively (19 Japanese subjects for speech, 7−13 German subjects for music motifs A and B)

After an inspection, the preferred delay is found roughly at a certain duration of autocorrelation function, defined by τ_p, such that the envelope of autocorrelation function becomes $0.1\,A_1$. Note that since the autocorrelation function is expressed with dimension of a power (see Eq. 2.3) it should be compared with a power amplitude A_1^2. However, the subjective results showed a better fit to the pressure amplitude A_1. The data collected are shown in Fig. 4.3 as a function of the duration τ_p. Thus, $\tau_p = \tau_e$ only when $A_1 = 1$. This means that the pressure amplitude of the reflection is ten times greater than the amplitude of the autocorrelation function at τ_p. Data from a continuous speech signal of $\tau_e = 12$ ms are also plotted in the figure.

Two reasons can be considered to explain why the preference decreases for $0 < \Delta t_1 < \tau_p$:

1) tone coloration effects occur because of the interference phenomenon in the coherent time region (Appendix C); and
2) the IACC increases, as shown in Fig. 3.5, for $\Delta t_1 \to 0$.

On the other hand, echo disturbance effects can be observed if Δt_1 is longer than τ_p [4.6], see Fig. 5.5a.

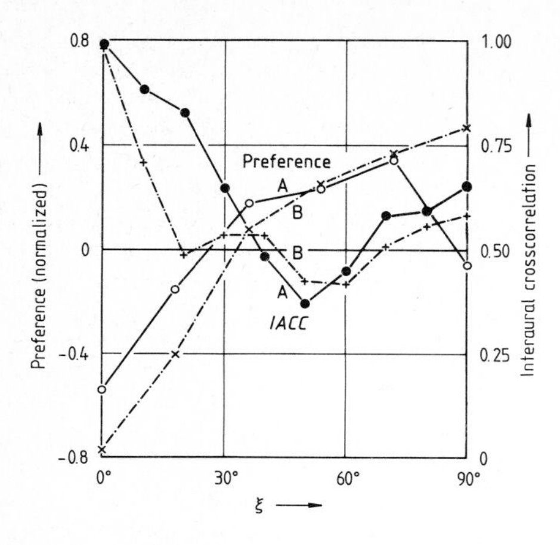

Fig. 4.4. Preference scores and the magnitude of IACC of the sound fields as a function of the horizontal angle of the single reflection, $A_1 = 0$ dB (6 sound fields and 13 German subjects) [3.1]

4.2.2 Preferred Direction of a Single Reflection

The delay time of the reflection, in an experiment to show the preferred direction of a single reflection, was fixed at 32 ms. The reflection was presented by loudspeakers located at $\xi = 18°$ ($\eta = 9°$), 36°, ..., 90° and $\xi = 0°$ ($\eta = 27°$).

Results of the preference tests are shown in Fig. 4.4 as a function of the echo direction ξ. The IACC is also plotted for comparison. No fundamental differences are observed between the curves of the motifs in spite of the great differences of source signal. The preference score increases roughly with decreasing IACC. The correlation coefficient between the score and the IACC is typically -0.8 (at 1% significance level). The score with motif A at $\xi = 90°$ drops to a negative value, indicating that the lateral reflections coming from around 90° are not always preferred. The figure shows that there is a preference for angles greater than $\xi = 30°$. On the average there may be an optimum range centered at 55°, say, $\pm(55° \pm 20°)$. Similar results can be seen in the data from speech signals [4.7].

4.2.3 Preferred Amplitude of a Single Reflection

Using music motif B, sound fields with a single reflection from the direction of $\xi_1 = 40°$ and $\eta_1 = 19°$ were examined. These values are in the range $\pm(55° \pm 20°)$. The angle measured from the median plane is $\alpha \approx 37°$ [$\alpha = \sin^{-1}(\sin\xi \cos\eta)$]. The delay and amplitude of the single reflection were chosen $\Delta t_1 = 10, 20, \ldots, 60$ ms and $A_1 = \pm 1.5$, ± 4.5, ± 9.5 dB, respectively.

First of all, to obtain the preferred delays at each fixed amplitude of reflection, paired comparison tests were performed by varying the delay for each fixed A_1. Results are shown in Fig. 4.5. Next, sound fields with the most preferred delay at each amplitude of reflection, $F_1, F_2, \ldots, F_6$, were systematically obtained. The preference scale values (linear) for all sound fields were cal-

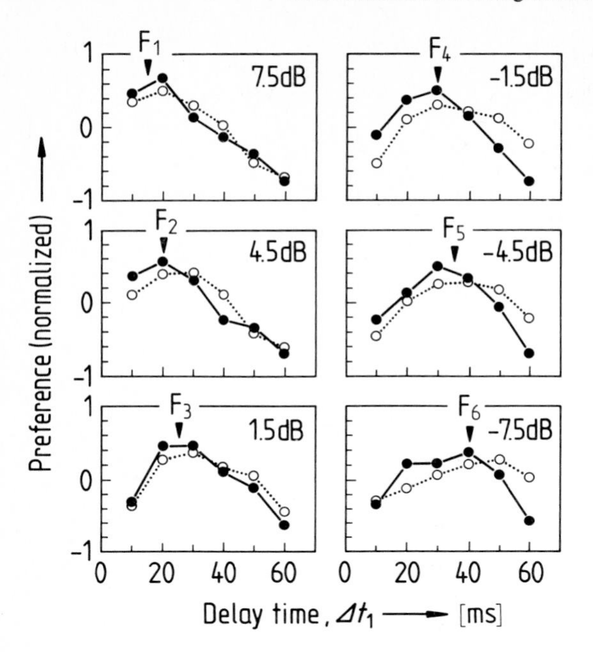

Fig. 4.5. Preference scores of the sound fields with single reflection as a function of its delay, for each fixed amplitude of the reflection (6 sound fields and 8 Japanese subjects) [4.8]. ●: Music piece of motif B: 0 – 5 s; ○: music piece of motif B: 5 – 15 s; (subscript) amplitude of reflection

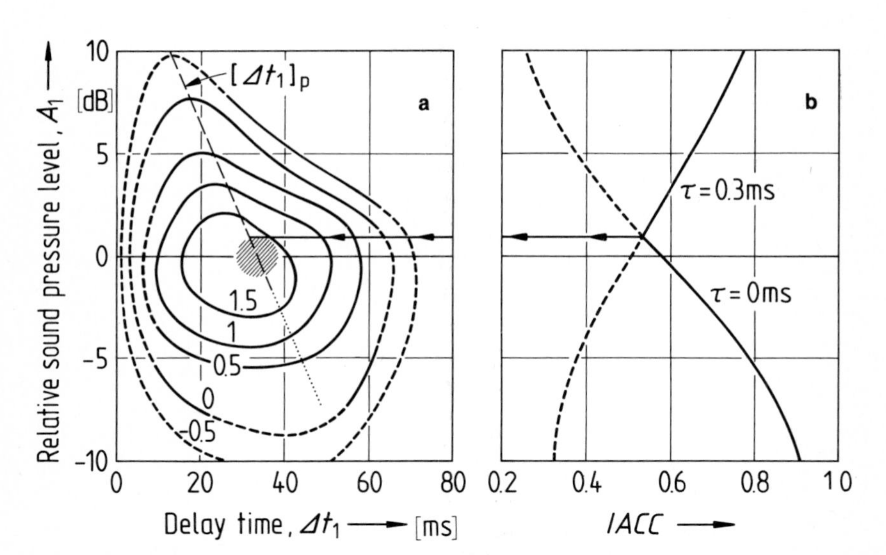

Fig. 4.6 a, b. Optimal conditions of the sound fields vs delay and amplitude of the single reflection, $\xi = 40°$, $\eta = 19°$ [4.8]. **(a)** Contour lines of equal preference. Scale values of preference are obtained by the law of comparative judgment, Sect. 4.1 (8 Japanese subjects); **(b)** Magnitude of IACC as a function of the amplitude of the single reflection. The minimum value of IACC is found at $A_1 \simeq +1$ dB, Fig. 4.5

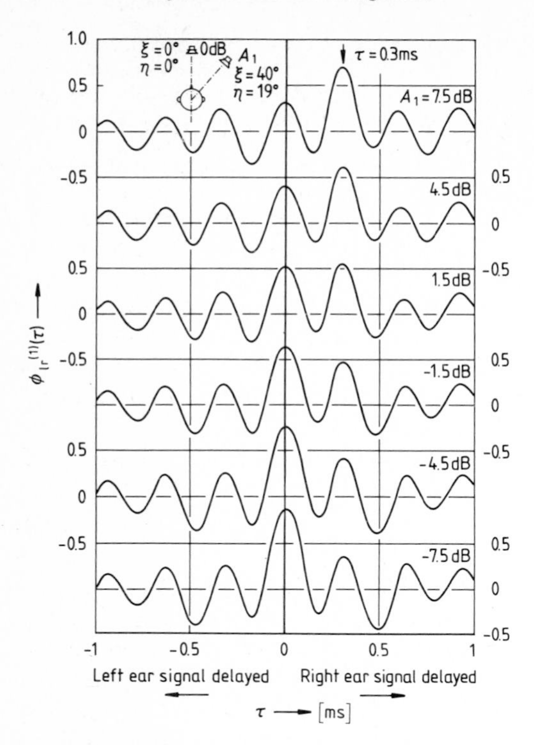

Fig. 4.7. Calculated values of the interaural cross correlation for sound fields with single reflection ($\xi = 40°$, $\eta = 19°$), as a parameter of the amplitude of the reflection (music motif B). Values of correlation functions needed for the calculation are indicated in Table D.1b

culated by assuming $\sigma_d = \sigma = 1$ in (4.3). Contour lines of the scale value are obtained by the proportional allotment as shown in Fig. 4.6a, as a function of the delay and amplitude of the reflection. Calculated values of the IACC as a function of the amplitude are shown together in Fig. 4.6b. The corresponding interaural cross correlations as a function of its delay for $|\tau| < 1$ ms and as a parameter of the amplitude A_1 are shown in Fig. 4.7. These values are obtained with (3.7) with the measured values listed in Table D.1b of correlations for a single sound. The maxima occur at $\tau = 0$ and $\tau = 0.3$ ms, corresponding to the directions of the direct sound and the reflection, respectively. The smallest magnitude of the IACC can be observed at $A_1 = +1$ dB. Under this condition, however, a spatial split of sound can be clearly perceived, because of the identical maxima of IACC at the delays $\tau = 0$ and $\tau = 0.3$ ms. Therefore, the optimal condition can be obtained at a lower amplitude, say, near $A_1 = 0$ dB, shown as a shaded region in Fig. 4.6a. The maxima of the interaural cross correlation should appear at $\tau = 0$, ensuring frontal localization of the continuous sound signal. The dashed line in Fig. 4.6a indicates the calculated preferred delay for each amplitude of reflection; the line is calculated with the autocorrelation function indicated in Fig. 2.2b, as described in Sect. 4.2.1. The preference data may well explain in terms of only the interaural cross correlation and the delay time. Therefore, the amplitude of reflection itself is unlikely to be an independent objective parameter to determine the preference.

4.2.4 Preferred Delay Time of the Second Reflection

Using a method similar to that described above, one can find the optimal conditions of sound fields with two early reflections. The directions of the reflections were fixed at $\xi_{1,2} = \pm 40°$ and $\eta_{1,2} = 19°$. At first, the delays between the first and second reflections were adjusted at $\Delta t_2 - \Delta t_1 = 10, 20, \ldots, 60$ ms (6 sound fields) for each fixed delay $\Delta t_1 = 10, 20, \ldots, 60$ ms. Next, to obtain the optimal conditions, the test was conducted with 9 sound fields in the combination of $\Delta t_1 = 20, 30, 40$ ms and $\Delta t_2 - \Delta t_1 = 10, 20, 30$ ms [4.8]. Therefore, the contour lines of equal preference can be drawn in the range indicated by solid lines in Fig. 4.8a, b for two amplitudes pattern of reflections, i.e., $A_1 = -4.2$ dB, and $A_2 = -6.2$ dB, and $A_1 = A_2 = 0$ dB, respectively. It is clear from both cases that the preferred initial time delay agrees with those calculated above, and described more generally by (5.4 – 6) below.

The preferred delay of the second reflection can be approximately found by

$$[\Delta t_2]_p \approx 1.8\,[\Delta t_1]_p\,. \tag{4.8}$$

Thus, in other words, $[\Delta t_2 - \Delta t_1]_p \approx 0.8\,[\Delta t_1]_p$, so that the second time gap between the first and second reflections should be shorter than the initial time-delay gap. Therefore, the preferred delay of the second reflection depends on that of the initial time-delay gap. This suggests that preferred delays of multiple early reflections are related to the preferred initial time-delay gap between the direct sound and the first reflection (0.8 ≈ root of the golden ratio!).

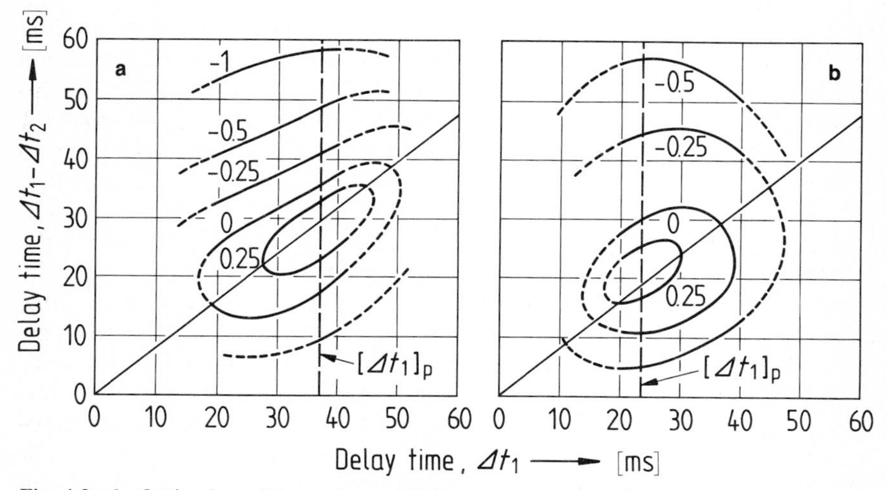

Fig. 4.8a, b. Optimal conditions of sound fields with two early reflections vs the two delay times (10 – 13 Japanese subjects) [4.8]. **(a)** Amplitudes of the two reflections: $A_1 = -4.2$ dB and $A_2 = -6.2$ dB; **(b)** amplitudes of the two reflections: $A_1 = A_2 = 0$ dB

4.2.5 Preferred Spectrum of a Single Reflection

Direct sound and a single reflection from the walls were simulated with the aid of a digital computer, using the transfer functions of the reflection from several boundary walls. The sound radiation from loudspeakers was corrected by the inverse filtering technique. The boundary walls used in simulating the sound fields were a perfect reflecting wall (a), a 5 cm thick glass-fiber wall (b), a perforated wall (c) and a periodic rib (d). Amplitudes of the transfer function of the walls are shown in Fig. 4.9. The transfer function of the glass-fiber wall, which behaves like a low-pass filter, was measured by the interference pattern method for the higher frequency range above 3.15 kHz [4.9] and was calculated assuming a locally reacting surface using the values measured by an acoustic tube method [4.10] for the lower frequency range. The transfer function of the perforated wall was measured by *Ingard* and *Bolt* [4.11]. The transfer function of the periodic rib wall is theoretically calculated at the position $y = 500$ cm from the surface (Sect. 2.3).

The impulse responses $w(t)$ obtained by the inverse Fourier transform of $W(\omega)$ are shown in Fig. 4.10 together with their wall structures. It is worth noting that durations of impulse response reach 2.0 ms except for that of the perfect wall, and that of the periodic rib wall which reached 12 ms. The direction of the reflection arriving at the listener was fixed at $\xi = 30°$ and $\eta = 0°$. Paired comparison tests of this section only were made to judge a preference between a sound field with only direct sound, and a sound field with direct and single reflection. The sound source is continuous (female) speech and its autocorrelation function is shown in Fig. 2.2e. Results of the percent preference for the perfect reflecting wall are shown in Fig. 4.11 as a parameter of the amplitude of reflection. If the preference is greater than 50%, then the sound fields with single reflection may be said to be preferred to the direct sound alone. The maximum preference of 100% occurs at $\Delta t_1 = 16$ ms ($A_1 = 0$ dB). The corresponding preferred delay calculated by (5.4–6) below is about 12 ms. When $A_1 = -6$ dB and $+6$ dB, the calculated values are about 16 ms and 8 ms, respectively. Initially, it was thought that speech processing is dif-

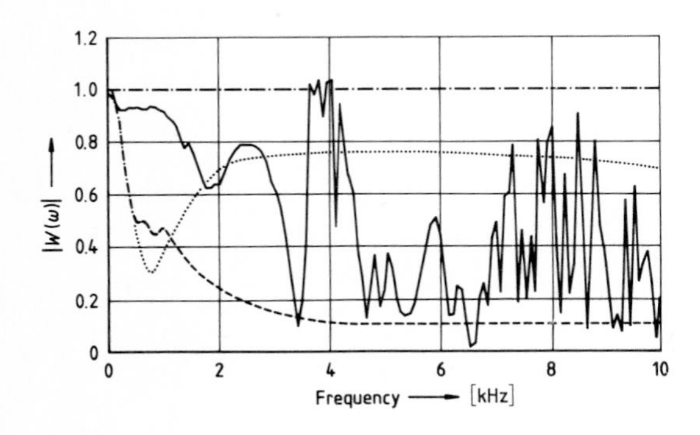

Fig. 4.9. Amplitudes of the transfer function for reflections from various boundary walls at oblique wave incidence upon the surface ($\theta = 60°$) [4.7].
(–·–·–) Perfect reflecting wall;
(– – –) glass-fiber wall;
(·········) perforated wall;
(———) periodic rib wall (receiving position $y = 500$ cm above the surface)

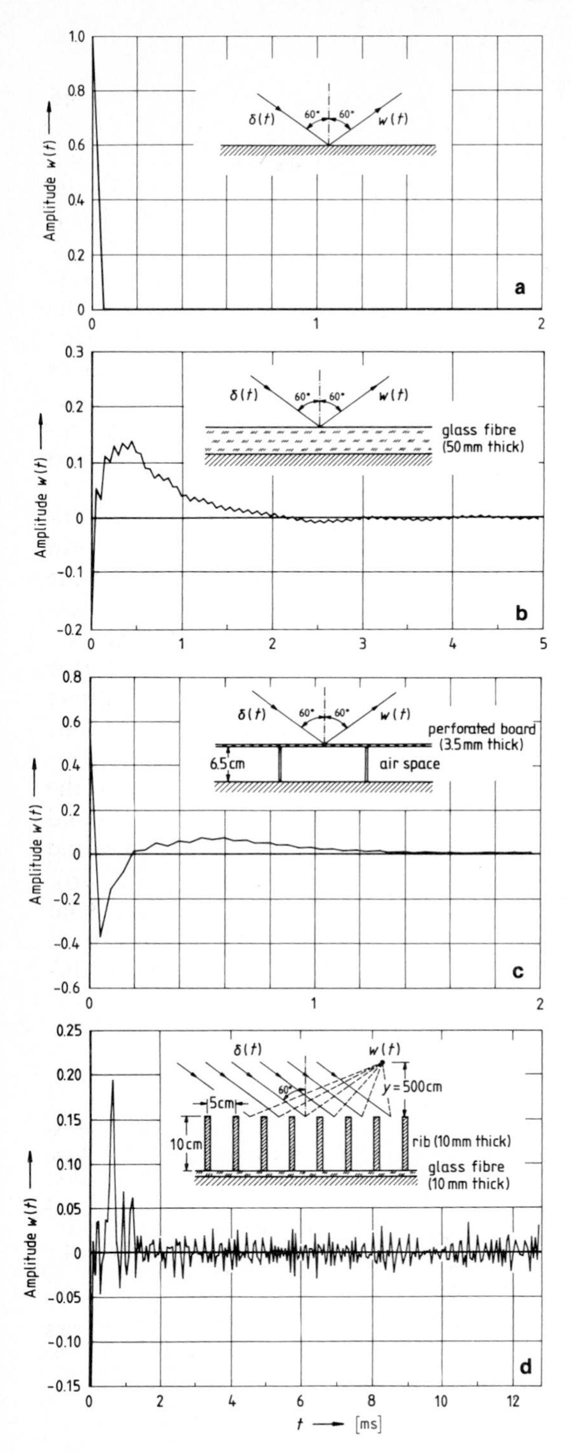

Fig. 4.10 a – d. Impulse responses for reflections from various boundary walls [4.7]. **(a)** Perfect reflecting wall; **(b)** glass-fiber wall; **(c)** perforated wall; **(d)** periodic rib wall (receiving position $y = 500$ cm above the surface)

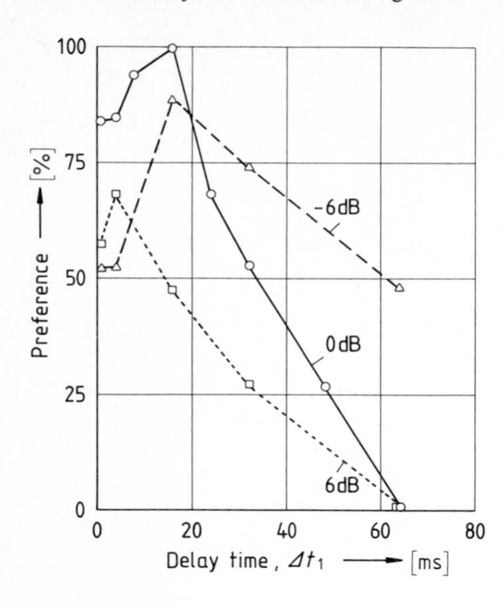

Fig. 4.11. Percent preference for the sound fields with a single reflection ($\xi = 30°$, $\eta = 0°$), as a function of delay and as a parameter of amplitude of the reflection (19 Japanese subjects) [4.7]

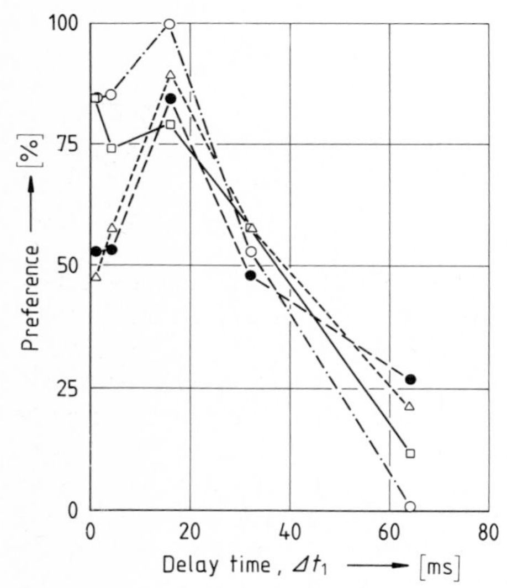

Fig. 4.12. Percent preference for sound fields with a single reflection ($\xi = 30°$, $\eta = 0°$) for various boundary walls, $A_1 = 0$ dB (19 Japanese subjects) [4.7].
(○ —·—·— ○) Perfect reflecting wall;
(○ ---------- ○) glass-fiber wall;
(△ ·········· △) perforated walls;
(□ ———— □) periodic rib wall (receiving position $y = 500$ cm above the surface)

ferent, but this indicates that the preferred delay for speech as well can be obtained in relation to the autocorrelation function.

To find preferred transfer functions when $A_1 = 0$ dB, results with different walls are shown in Fig. 4.12. The best preference is obtained for the perfectly reflecting wall. When the spectra of reflecting sound differ from those of direct sound, the preference value at each maximum is decreased. This effect has been explained by *Békésy* as funneling of the loudness of the reflected sound into that of the direct sound. Funneling requires that the frequency spectrum of the reflected sound not be different from that of the direct sound, so that the two sounds may readily mix and combine their loudness [4.12]. It is noteworthy in Fig. 4.9 that the transfer function amplitudes of walls b, c and d are usually less than 0.8 or fluctuate in the frequency range 0.4 – 3.5 kHz which mainly contains speech spectra.

4.2.6 Preferred Delay Time of the Strongest Reflection in Multiple Early Reflections

The set of experiments was extended to sound fields with four early reflections [4.13]. The tests were conducted with the ampitude decay of the reflections, the initial time-delay gap and the IACC as parameters. The results showed that sound fields with a smaller IACC were always preferred. The most preferred delay time of the strongest reflection can be determined by the autocorrelation function of source signals and the total amplitude of the reflections, as expressed by (5.4 – 6) below. Therefore, the delay time of the first reflection becomes unimportant if it is not the strongest reflection.

4.3 Sound Fields with Early Reflections and Subsequent Reverberation

All the significant objective parameters used to describe the sound signals at both ears in a concert hall can be reduced into the following: (1) the level of listening, (2) the delay time of early reflections, (3) the subsequent reverberation time and (4) the IACC.

To examine the independence of the effects of the objective parameters on the subjective preference judgments and to make continuation the scale value of preference to any sound field, two of four parameters were simultaneously varied while the remaining two were held constant. First, sound fields in a concert hall were simulated by computer with various combinations of early reflections and subsequent reverberation after the early reflections [4.14]. Second, sound fields with various combinations of the listening level and the IACC were simulated [4.15]. Third, sound fields with various combinations of subsequent reverberation time and the IACC were simulated [4.16]. In addition, results of previous preference judgments are also described.

4.3.1 Scale Values vs Delay Time of Early Reflections and Subsequent Reverberation Time

For convenience' sake, the sound fields in a concert hall with the same plan as the Symphony Hall in Boston (Fig. 4.13) were simulated. A modified system for simulating the sound fields in concert halls is shown in Fig. 4.14. By using this system, sound fields may be simulated for subjects with different head-related transfer functions (Sect. 3.2). Note that the system indicated in Fig. 3.6

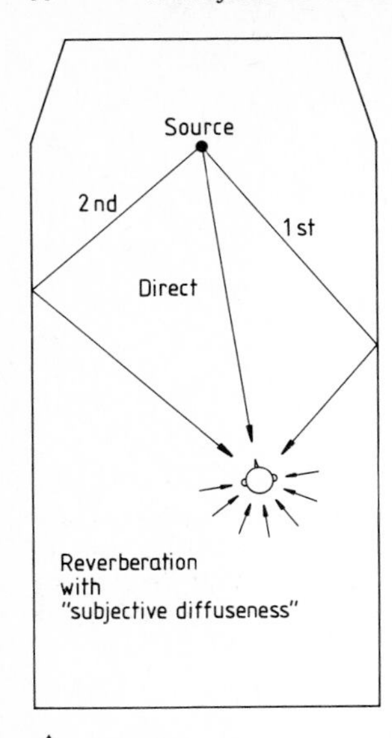

▲ **Fig. 4.13.** A sound field in a concert hall with the same plan as Symphony Hall in Boston simulated for direct sound, two early reflections and subsequent reverberation with "subjective diffuseness" [4.14]

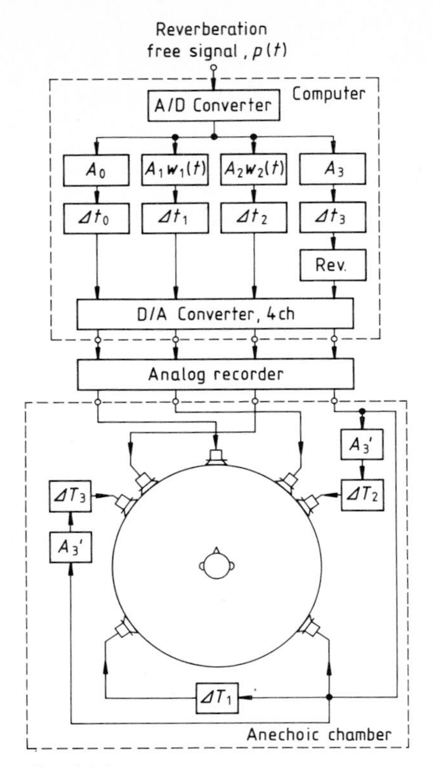

Fig. 4.14

Fig. 4.14. Simulation system of a sound field. Directions of arrival of the direct sound and two early reflections at the listener coincided with those drawn in Fig. 4.13. Directions of four loudspeakers for subsequent reverberation (Rev.) were chosen within $\pm(55° \pm 20°)$ from the median plane. Reverberation signals supplied to the loudspeakers were delayed by ΔT_j, $j = 1, 2, 3$. These produced a small magnitude of IACC

required the individual head-related transfer function for simulating sound localization. The computer program provides the time delay Δt_n of the early reflections and the subsequent reverberation relative to the direct sound. The reverberation signal, which was independent of frequency, was generated by the Schroeder reverberator, Sect. 3.3. To obtain a natural sound, the values of delay given in Fig. 3.10 were chosen as

$$\tau_i = (23.5, 25.2, 27.0, 30.5, 7.3, 2.1) \times T_{\text{sub}} \quad [\text{ms}] ,$$
$$i = 1, 2, 3, 4, a, b \quad \text{and} \tag{4.9}$$
$$g_i = (0.85, 0.84, 0.83, 0.81, 0.70, 0.70) ,$$
$$i = 1, 2, 3, 4, a, b ,$$

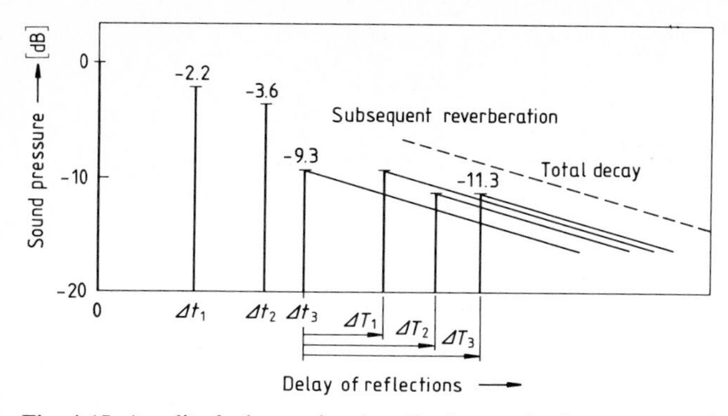

Fig. 4.15. Amplitude decay of early reflections and subsequent reverberation or impulse response of the sound fields tested [4.14]. Ampitudes of the two early reflections were determined by the $(1/r)$ law for the paths from the source point to the listener with reference to that of direct sound. Here T_{sub} is defined as a time interval in which the reverberation level drops by 60 dB after the early reflections

where T_{sub} is a reverberation time [s] or a decay time to decrease by 60 dB after the early reflections. Thus, the subsequent reverberation times were controlled by adjusting only the delays τ_i, $i = 1-4$, with colorless conditions of the all-pass filter, $i =$ a, b. To produce low magnitudes of the IACC in the sound fields for the source signals [3.1, 3] and to fix the spatial-binaural criterion, $h_{nl,r}(t)$, $n = 0, 1, 2, \ldots, 6$, signals with discrete time delays were fed to the loudspeakers (No. 1, 2, ..., 6) fixed at azimuth angles, i.e., $\xi_{1,2} = \pm 45°$ $(\pm 5°)$ (the elevation angle $\eta = 0°$), $\xi_{3,4} = \pm 125°$ $(\eta = 0°)$ and $\xi_{5,6} = \pm 55°$ $(\eta = 15°)$, respectively. The amplitude decays of the early reflections and the subsequent reverberation are shown in Fig. 4.15. The pressures at both ear entrances of the listener in an anechoic chamber are expressed by

$$\begin{aligned} f_{l,r}(t) = {} & p_s(t) * h_{0l,r}(t) \\ & + p_s(t) * A_1 \delta(t - \Delta t_1) * h_{1l,r}(t) \\ & + p_s(t) * A_2 \delta(t - \Delta t_2) * h_{2l,r}(t) \\ & + \sum_{n=3}^{N} [p_s(t) * A_n \delta(t - \Delta t_n) * h_{3l,r}(t) \\ & + p_s(t) * A_n \delta(t - \Delta t_n - \Delta T_1) * h_{4l,r}(t) \\ & + p_s(t) * A_3' A_n \delta(t - \Delta t_n - \Delta T_2) * h_{5l,r}(t) \\ & + p_s(t) * A_3' A_n \delta(t - \Delta t_n - \Delta T_3) * h_{6l,r}(t)] \, , \end{aligned} \tag{4.10}$$

where $p_s(t) = p(t) * s(t)$, $s(t)$ is the radiation impulse response of the loudspeakers. The amplitude response of the loudspeakers for the frequency range 110 Hz – 8.5 kHz was ± 3.5 dB, and the difference between loudspeakers was within 1 dB. Also, $A_0 = 0$, $A_1 = -2.2$, $A_2 = -3.6$, $A_3 = -9.3$, $A_3 A_3' = -11.3$, $A_n \geqq -49.3$ and $A_3' A_n \geqq -51.3$ when $n \leqq N$, $A_n = A_3' A_n = -\infty$, $n \geqq N$ [dB]; and

$$\Delta t_1 = 22 \times \mathrm{SD}\,, \quad \Delta t_2 = 38 \times \mathrm{SD}\,, \quad \Delta t_3 = 47 \times \mathrm{SD}\,,$$

SD being the scale of dimension of the auditoriums: furthermore,

$$\Delta T_1 = 18\,, \quad \Delta T_2 = 30\,, \quad \Delta T_3 = 40 \quad [\mathrm{ms}]\,,$$

making the four reverberation signals incoherent (Fig. 3.5). Consequently, the amplitude of the total reverberation decayed smoothly after the two early reflections until it dropped by about 50 dB. The time delay $\Delta t_2 - \Delta t_1$ between the first and second reflections was chosen as $0.73\,\Delta t_1$, within the preferable conditions (Sect. 4.2.4).

Paired comparison tests of the 16 sound fields for each source signal were conducted for changes in the temporal-monaural criteria only, i.e., the scale factor of the early reflections SD and the subsequent reverberation time T_{sub}. To change the autocorrelation function, several types of source signals were used, i.e., motifs A, B and E, also speech signals. The long-time autocorrelation function ($2T = 35$ s) of each source signal was measured with an A-weighting network after it passed through the loudspeaker with impulse response $s(t)$, which was used for producing the direct sound in the preference tests. The autocorrelation functions τ_e were slightly different as described in Table 2.1 because of different radiation characteristics of the loudspeaker used. To present the preferred loudness for each source signal, the total sound pressure levels in the sound fields were adjusted to each preferred listening level, Sect. 4.2.3. Five-second passages of source signals, except for 10 s of motif A, were used for the paired comparison tests. Test signals were presented with a 1 s interval between the pairs. Each subject (13 subjects for motif A and 14 subjects for the other signals, 23 ± 3 years old) judged which of the sound fields they preferred to hear in a concert hall or in a lecture room. The tests were performed for all combinations of the pairs, i.e., 240 pairs for each source signal, interchanging the order in each pair. Since they took about five hours in total for each subject, sessions were limited to 12 minutes to avoid fatigue effects. Scale values of preference, which are regarded as a linear psychological distance between the sound fields, were obtained by applying the law of compartive judgment (case V) [4.1], and were confirmed by the test of goodness of fit throughout the investigation [4.4].

The scale values of preference for the 16 sound fields are indicated in Table 4.1a, b. The agreement of judgments between the subjects for each source signal was sufficient; responses of the subjects were not random (according to the chi-square test, at 5% significance level). Contour lines of equal preference were drawn by the proportional allotment, shown in Fig. 4.16, as a function of Δt_1 (or SD) and T_{sub}. It is obvious that the optimal conditions differ greatly between source signals. The conditions can be easily found at:

$\Delta t_1 \simeq 70$ ms (SD $\simeq 3.5$) and $T_{sub} \simeq 2.6$ s for motif A (Gibbons), and

$\Delta t_1 \leqq 10$ ms (SD $\leqq 0.5$) and $T_{sub} \simeq 0.9$ s for motif B (Arnold).

A tendency is clear; contour lines drawn at every 0.25 interval become more dense with decreasing coherence of the autocorrelation function. The most

Table 4.1. Scale values of preference for the 16 sound fields

a) Music motif A (9 subjects) $\tau_e = 127$ ms

	T_{sub} [s]			
SD	1.5	3.0	4.5	6.0
1.0	0.19	0.23	0.22	−0.13
3.0	0.26	0.37	0.07	−0.37
5.0	0.20	0.20	−0.31	−0.33
7.0	0.07	0.06	−0.27	−0.45

b) Music motif B (14 subjects) $\tau_e = 43$ ms

	T_{sub} [s]			
SD	0.5	1.0	1.5	2.5
0.2	0.90	1.10	0.11	−0.79
0.8	0.91	0.81	−0.16	−0.99
1.4	0.53	0.52	−0.21	−1.11
2.0	0.05	0.08	−0.40	−1.36

Table 4.2. Analyses of variance for three tests A, B, and C. Reliabilities (95%) of scale values are obtained less than ±0.16 (Music motif A) and ±0.21 (Music motif B)

Test	Factor	Sum of squares	Degree of freedom	Mean square	F	Significance level	Contribution [%]
a) Music motif A							
	Δt_1 (SD)	0.20	3	0.07	4.4	<0.05	14
A	T_{sub}	0.73	3	0.24	17	<0.01	65
	Residual	0.13	9	0.01	–		
	L. Level	0.99	3	0.33	48	<0.01	27
B	IACC	2.61	2	1.30	187	<0.01	71
	Residual	0.04	6	0.01	–		
	T_{sub}	2.44	3	0.82	68	<0.01	89
C	IACC	0.17	3	0.06	5	<0.05	5
	Residual	0.11	9	0.01	–		
b) Music motif B							
	Δt_1 (SD)	1.20	3	0.40	22	<0.01	13
A	T_{sub}	7.63	3	2.54	141	<0.01	84
	Residual	0.16	9	0.02	–		
	L. Level	0.74	3	0.25	12	<0.01	24
B	IACC	1.90	2	0.95	47	<0.01	67
	Residual	0.12	6	0.02	–		
	T_{sub}	2.55	3	0.85	182	<0.01	79
C	IACC	0.64	3	0.21	46	<0.01	19
	Residual	0.04	9	0.01	–		

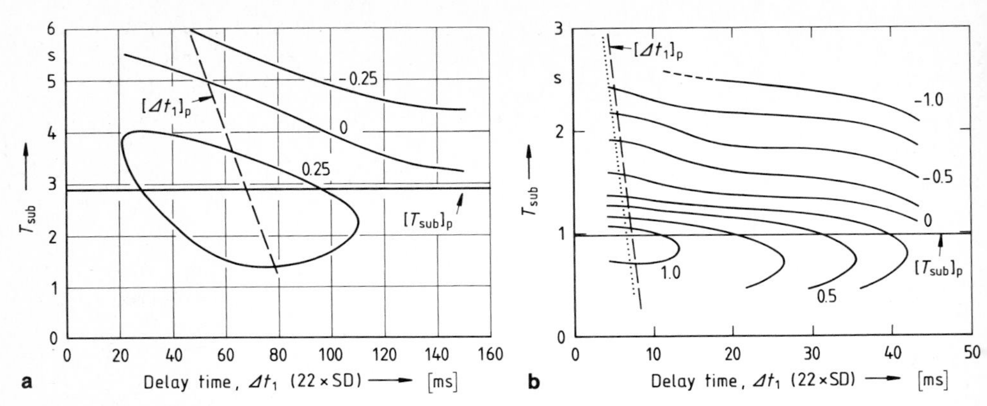

Fig. 4.16 a, b. Contour lines of equal preference for the sound fields vs Δt_1 (or SD) and T_{sub} [4.14] using the values in Table 4.1 (16 sound fields). The dashed line is the preferred value of the initial time-delay gap between the direct sound and the first reflection, calculated by (5.4–6). The solid line is the preferred value of the subsequent reverberation time, calculated by (4.11). **(a)** Music motif A (Gibbons), (9 Japanese subjects); **(b)** music motif B (Arnold), (14 Japanese subjects)

preferred initial time-delay gap $[\Delta t_1]_p$ is calculated similarly as before (Sect. 4.2 and (5.4–6)). Calculated values of $[\Delta t_1]_p$ preferred initial time-delay gaps, are shown by the dashed lines in Fig. 4.16.

The results of the analysis of variance with the scale values for each source signal are indicated in Table 4.2 (Test A). Factors Δt_1 (SD) and T_{sub} are reasonably independent of each other because of the small residuals. Also, contributions of factor T_{sub} are superior to those of Δt_1 (SD) in the range tested.

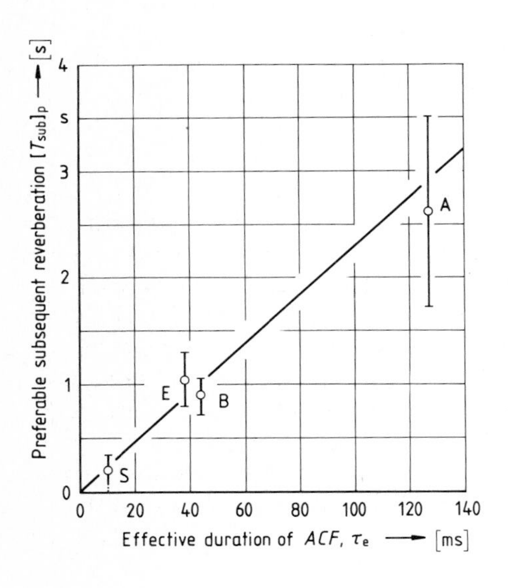

Fig. 4.17. Relationship between the preferable subsequent reverberation time and the effective duration of autocorrelation function, i.e., τ_e such that $|\phi_p(\tau)|_{envelope} = 0.1$. Ranges of preferable reverberation are graphically obtained at a 0.1 below the maximum scale values of preference [4.14]. A, B, E and S refer to the source signals

Figure 4.17 shows the relationship between preferred subsequent reverberation time and the effective duration of the autocorrelation function. The statistical correlation between them is 0.99 (at 1% significance level). The preferred subsequent reverberation time $[T_{sub}]_p$ is tentatively described by

$$[T_{sub}]_p \simeq 23\ \tau_e\,[s]\ . \quad (4.11)$$

For example, it is 2.9 s (23 × 0.127 s) for music motif A. The solid lines in Fig. 4.16 are calculated values. As is clear from these results, the preferred reverberation time is nearly independent of Δt_1 or SD, in spite of the two parameters belonging to the temporal-monaural factor.

So far, the total amplitude of reflections was set at $A \approx 4.1$, because the sound fields at a rear position in a concert hall were simulated. When we examined the sound field at a front seat in the hall with $A \approx 1.1$, the preference results showed that a longer reverberation time was acceptable. However, the most preferred reverberation time agrees fairly well with the calculation results of (4.11), as shown in Fig. 5.9. Note that the total amplitude of reflections is related to "Deutlichkeit" as defined by Thiele, Sect. 7.2.

4.3.2 Scale Values vs Listening Level and IACC

Next we examined how the subjective preference is independent of the listening level and the interaural cross correlation. The sound fields were simulated with the same system described above. To obtain the optimum conditions, other parameters were fixed at the preferred values. Accordingly, the subsequent reverberation times were kept constant: 3.0 s for music motif A and 1.0 s for music motif B.

The initial time-delay gaps between the direct sound and the first reflection were kept at 80 ms for music motif A and 20 ms for music motif B respectively. These values were calculated by (5.4, 6) with A = 2.0. Also, the time delays of reflections were kept at $\Delta t_2 - \Delta t_1 = 0.8\ [\Delta t_1]_p$ and $\Delta t_3 - \Delta t_2 = 0.64\ [\Delta t_1]_p$. The values of IACC were adjusted in the range of 0.4 – 1.0 by changing the direction of loudspeakers which produce both early reflections and subsequent reverberation. The loudspeaker arrangements are shown in Fig. 4.18. Under these conditions, the maximum value of the interaural cross correlation was always maintained at $\tau = 0$. Loudspeaker system (a) included the reflections with the optimum angles $\xi = \pm 55°$ (IACC ≈ 0.4), while system (b) contained early reflections from the listener's median plane and from certain angles in the horizontal plane (IACC ≈ 0.7). System (c) included only reflections from the median plane (IACC = 0.98: absorbing side walls).

The preference judgments were conducted for the 12 fields with 16 subjects. Most subjects judged all sound fields in the above two experiments. The contour lines of equal scale value of preference are shown in Fig. 4.19, with

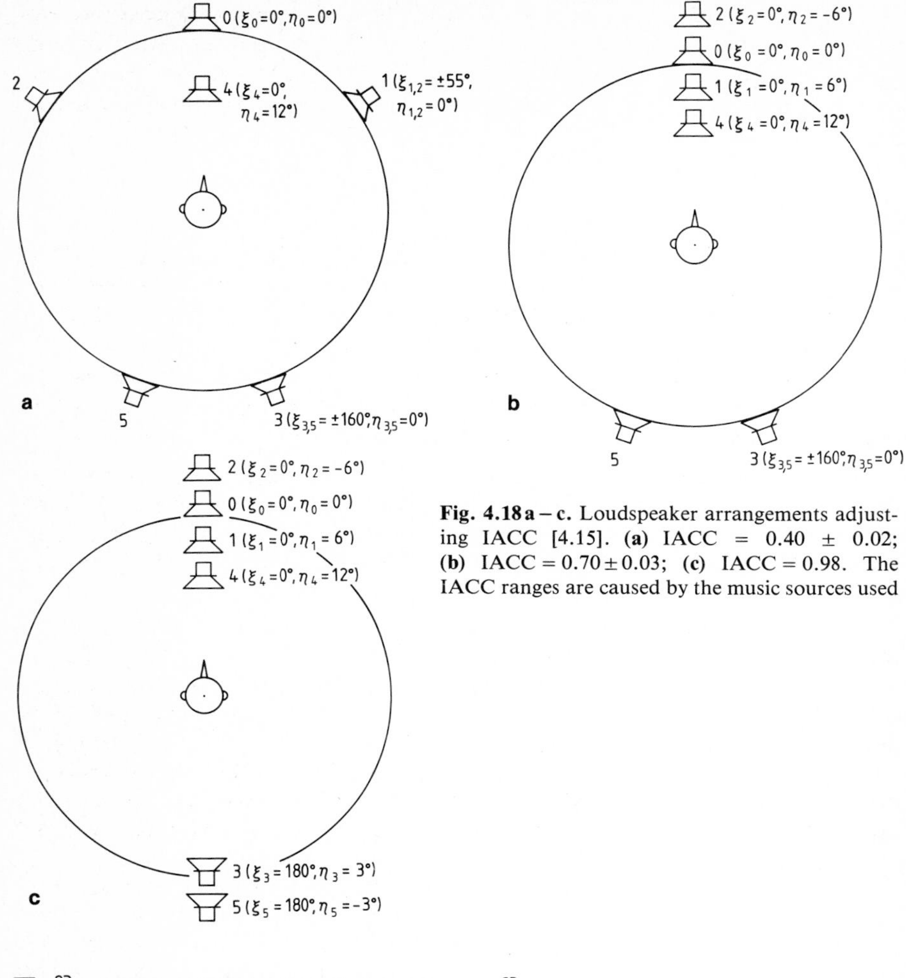

Fig. 4.18 a – c. Loudspeaker arrangements adjusting IACC [4.15]. **(a)** IACC = 0.40 ± 0.02; **(b)** IACC = 0.70 ± 0.03; **(c)** IACC = 0.98. The IACC ranges are caused by the music sources used

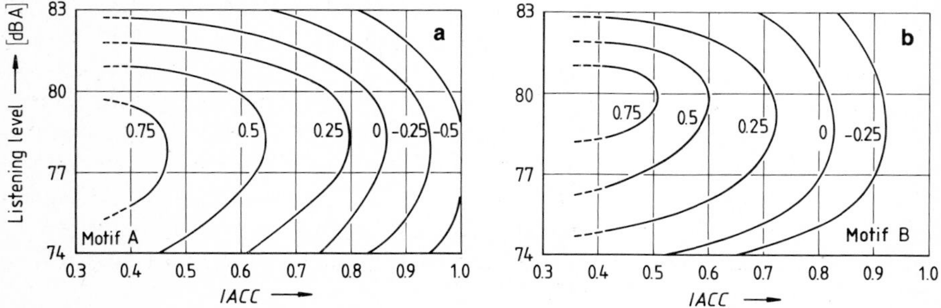

Fig. 4.19 a, b. Contour lines of equal preference for the sound fields vs IACC and listening level (12 sound fields) [4.15], drawn using the values in Table 4.3. **(a)** Music motif A (Gibbons), (16 Japanese subjects); **(b)** music motif B (Arnold), (16 Japanese subjects)

Table 4.3. Scale values of preference for each sound field obtained by paired comparison tests (16 subjects, 12 sound fields)

a) Music motif A

System	IACC	Listening level [dBA][a]			
		74	77	80	83
(a)	0.39	0.53	0.85	0.73	−0.07
(b)	0.72	0.07	0.35	0.36	−0.35
(c)	0.98	−0.64	−0.35	−0.50	−0.98

b) Music motif B

System	IACC	Listening level [dBA][a]			
		74	77	80	83
(a)	0.42	0.10	0.55	0.92	0.20
(b)	0.67	−0.35	0.27	0.37	0.04
(c)	0.98	−0.85	−0.38	−0.41	−0.46

[a] The maximum amplitudes were measured with a high damping of the "slow" meter (B&K Type 2203)

values taken from Table 4.3. For a given constant listening level, the sound fields with a smaller IACC are always preferred.

The preferred listening levels depend upon the source music presented, but they are hardly affected by the IACC. The preferred listening levels are found in the ranges 77 – 79 dBA for music motif A and 79 – 80 dBA for music motif B respectively. Note that these results are not related to "musicality", because of the short music pieces used.

Results of the analyses of variance with the scale values of preference for each motif are indicated in Table 4.2 (Test B). It is clear that the two parameters (the listening level and the IACC) are independent influences on the subjective preference judgments. Also, contributions of the IACC to changes in preference are larger than those of the listening level in the ranges tested.

Since the scale values of preference are derived as a linear psychological distance between sound fields, we can superpose the scale value for each parameter, if the parameters are independent. We wish now to confirm this principle by an example. The average scale values of preference with 95% reliabil-

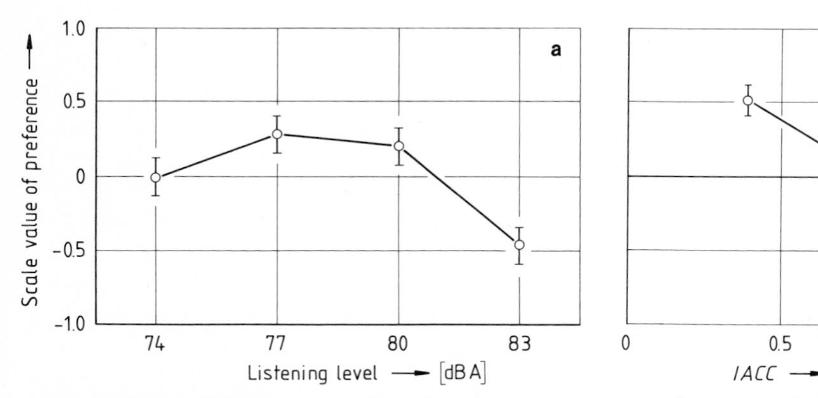

Fig. 4.20a, b. Average scale values of preference with 95% reliability (music motif A) [4.15] (**a**) for listening level; (**b**) for IACC

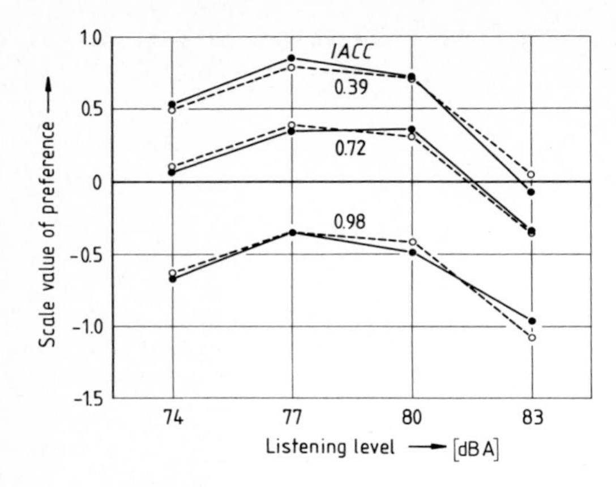

Fig. 4.21. Scale values of preference for each sound field as a function of listening level and as a parameter of IACC (music motif A) [4.15]. (● —— ●) Scale values obtained by preference tests; (○ - - - ○) scale values calculated with (4.12) with average scale values shown in Fig. 4.20

ity as a function of the listening level and of the IACC are shown in Fig. 4.20a, b, respectively. The scale values of preference for each sound field with any combination of the parameters can be obtained by adding the average values, so that

$$S = S\,(\text{Level}) + S\,(\text{IACC})\,, \tag{4.12}$$

where S (Level) and S (IACC) are the scale values as a function of the listening level and the IACC, respectively. As shown in Fig. 4.21 observed and calculated values closely agree. Similar results have been obtained for music motif B as well [4.15]. Thus, the principle of superposition may be applied for the linear scale values of preference if the parameters are independent.

4.3.3 Scale Values vs Subsequent Reverberation Time and IACC

In the third step of one experimental set, the independent effect of the subsequent reverberation time and the interaural cross correlation on subjective preference were examined. Also observed was the consistency of preference scale values from the different series of preference tests. As before, the other two parameters were fixed at the preferred conditions. The values of IACC were adjusted in the range 0.31 – 0.93 by changing the direction of loudspeakers located around the listener. The values of the reverberation time were adjusted in the range 1.5 – 6.0 s for music motif A and 0.5 – 2.5 s for music motif B. The preference judgments were conducted for 16 sound fields with 8 subjects.

The results of analyses of variance confirm that the reverberation time and the IACC independently influence subjective preference (Table 4.2, Test C). The most preferred reverberation times obtained here confirm (4.11) [4.16].

As we have seen, the preferred condition can generally be obtained by minimizing the magnitude of the IACC. However, the maximum value of the cross correlation must be maintained at $\tau = 0$ to ensure frontal localization of the sound source. The preference data satisfied by this condition is described in Sect. 5.3 (Figs. 5.8, 10), demonstrating a reasonable degree of consistency in the scale values of preference.

4.3.4 Agreement with Other Preference Judgments

In addition, subjective preference judgments were performed in sound fields with four multiple early reflections [4.13] and with two early reflections and subsequent reverberation time [4.17]. These results confirm that the most preferred initial time-delay gap can be found by (5.4, 6) and also that factors Δt_1 (SD) and the IACC are independent of each other in subjective preference judgments.

To obtain a total scale value of preference for any sound field with independent parameters, a consistent of unit of the scale values obtained from these preference judgments will be discussed in Sect. 5.3.

5. Prediction of Subjective Preference in Concert Halls

The purpose of this chapter is to evaluate acoustic quality at each seat of concert hall prior to its final architectural design. First of all, a workable model of the auditory system is presented which conforms to the auditory physiological system as described by the present level of knowledge. Then, the optimal design objectives and the linear scale value of preference for each parameter are given. Finally, as an example, the scale values of preference at each seat in a concert hall are calculated and the contour lines of equal preference are illustrated in the plan of the hall.

5.1 Model of Auditory Pathways

Both an auditory pathway model as well as the sound transmission system from the sound source to both ears are shown in Fig. 5.1. Based on the subjective preference judgments discussed in the previous chapter, correlation models of the auditory pathways are proposed in this chapter. The power density spectra in the neural activities in both auditory pathways are probably transformed into the autocorrelation functions $\Phi_{ll}(\sigma)$ and $\Phi_{rr}(\sigma)$, respectively, where σ corresponds to the position of neural activities. The power density spectrum of the direct sound may be separated in the neural processing by focusing on the "target signal" arriving from the front, and then from there the spectrum is transformed into the autocorrelation function of the source signal $\Phi_p(\sigma)$ which includes the effect of ear sensitivity. These transformations are performed, it is supposed, in a manner equivalent to the Fourier cosine transform. After acquiring the cross-power density spectra, including interaural time-delay characteristics, the interaural cross correlation may be neurally attained by the inverse Fourier transform.

Another possibility is that the interaural cross correlation $\Phi_{lr}(\sigma)$ in the "time domain" is directly derived by the neural activity above, associated with the autocorrelation function, so that

$$\Phi_{lr}(\sigma) \approx \Phi_{h_l h_r}(\sigma) * \Phi_p(\sigma)\,, \qquad \text{where} \tag{5.1}$$

$$\Phi_{h_l h_r}(\sigma) = \int_{-\infty}^{+\infty} h_l(r|r_0;t)\, h_r(r|r_0;t+\sigma)\, dt\,. \tag{5.2}$$

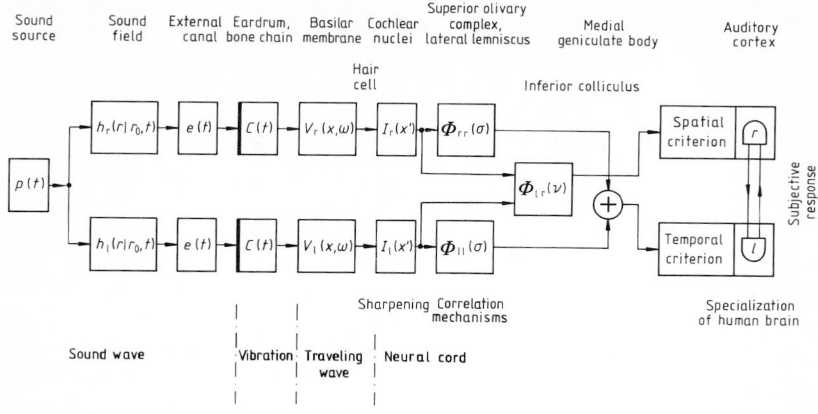

Fig. 5.1. Model of processors in the auditory pathways for subjective preference judgments: $p(t)$: source signal in the time domain; $h(r|r_0, t)$: impulse response between the source point r_0 and the ear entrance of a listener sitting at r; $e(t)$: impulse response between ear entrance and eardrum; $c(t)$: impulse response of the bone chain between eardrum and oval window including the transformation factor into vibration motion at the eardrum; $V(x, \omega)$: wave form of the basilar membrane, where x is the position along the basilar membrane measured from the oval window: $\Phi_{ll}(\sigma)$ and $\Phi_{rr}(\sigma)$: autocorrelation processors; $\Phi_{lr}(\nu)$: interaural cross-correlation processor; $\oplus$: signals combined. Output signals of a spatial cross-correlation mechanism and temporal autocorrelation mechanisms are assumed to be interpreted by different hemispheres, labeled r and l, respectively

Comparison between the sound field and the target alone may build up a neural formation corresponding to (5.2) for the spatial-environment perception. Though the operation seems to be a laborious task for the nervous system, the operation may be assumed to be performed in the "time domain" because the maximum interaural time-delay is limited to within about 750 μs. Several other possible mechanisms have been proposed for the correlation processors in the nervous systems [5.1 – 3].

The interaural cross correlation may be characterized by the cross correlation of the impulse responses from the sound source to both ears. Thus, for given positions of sound source and seat, the value of the IACC defined by (3.8) does not differ greatly from source signal to another (Appendix D).

The total amplitude of reflections, given by (5.6) below, is assumed to be determined in the neural pathway in the following manner:so that

$$A^2 = \frac{[\Phi_{ll}(0) + \Phi_{rr}(0)] - 2\,\Phi_p(0)}{2\,\Phi_p(0)}, \tag{5.3}$$

where $\Phi_{ll}(0)$ and $\Phi_{rr}(0)$ are the total intensities at each ear, and $\Phi_p(0)$ is the intensity of the direct sound only.

The autocorrelation mechanism is also a temporal criterion for the sensations of coloration (Appendix C), masking of echoes, loudness and perception of speech.

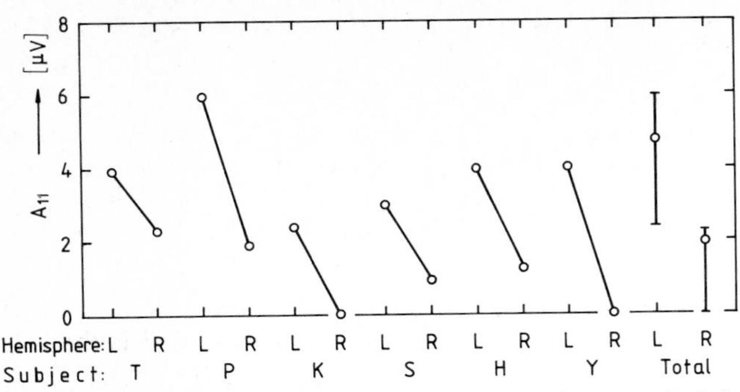

Fig. 5.2. Amplitude differences between averaged AEPs over the left (L) and the right (R) hemispheres, when a 90 ms natural vowel /a/ without any spatial variation was reproduced with a 300 ms interval (after Kikuchi [5.5])

In Chap. 4, it was pointed out that the spatial parameter (IACC) influences the preference judgments independently of the temporal parameters. It is quite natural to assume that processing of spatial and temporal factors are separately performed in different cerebral hemispheres. *Altman* et al. [5.4] applied the method of unilateral electroshock therapy in psychiatry in which a temporal functional disorder of the stimulated cerebral hemisphere results. They suggested that there was less confusion in determining sound localization in the right hemisphere than in the left. The right hemisphere is presumed to be concerned with spatial orientation. On the other hand, *Kikuchi* [5.5] has found, by means of recording auditory evoked potentials (AEP), left hemisphere dominance under presentation of vowel /a/, as shown in Fig. 5.2.

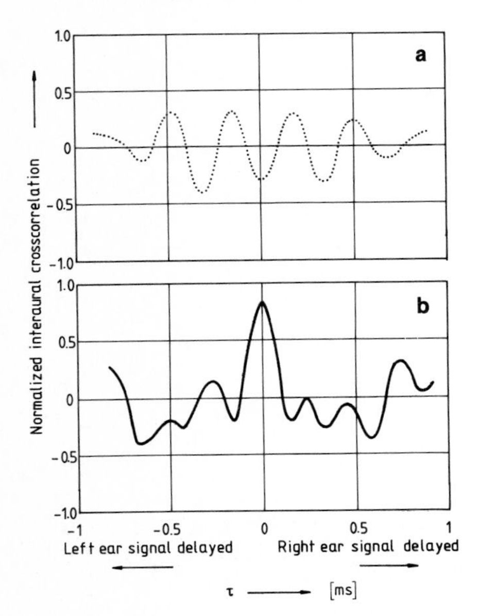

Fig. 5.3a, b. Measured interaural cross correlations of the two different spatial sound stimuli. **(a)** Stimulus A, "diffuse", **(b)** Stimulus B, "frontal"

In order to examine the hemispheric difference with respect to the appreciation of auditory spatial sensation, dynamic differences were examined through AEPs when a series of spatial sound stimuli with the vowel /a/ and changing IACC were presented [5.6]. This method enables us to precisely compare dynamic responses in AEPs over both hemispheres of each stimulus and is meaningful in connection with the results obtained from the paired comparison tests of subjective preference.

Measured interaural cross correlation functions of the vowel /a/ are shown in Fig. 5.3. The stimulus A is the low IACC value (0.41) of the sound field with clearly perceived subjective diffuseness. The stimulus B is the high IACC value (0.83) with perception of a frontal source direction which corresponds to a peak at $\tau = 0$. Since the chosen source signal is a part of continuous speech vowel /a/, even the running interaural cross correlation is invariable. The stimuli were 100 ms in duration without any click and the sound

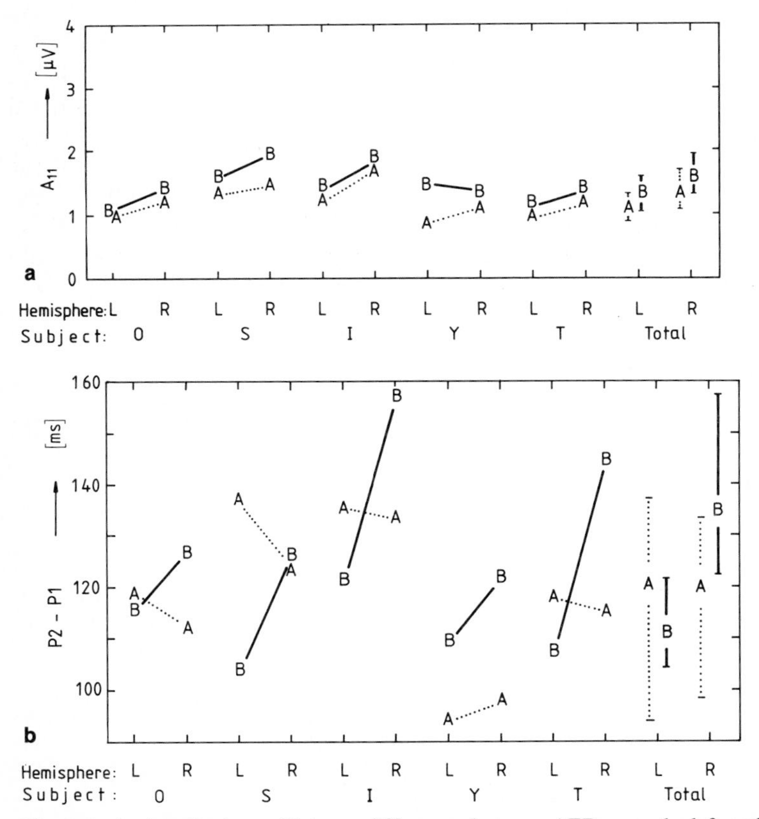

Fig. 5.4a, b. Amplitudes and latency-differences between AEPs over the left and the right hemispheres under dynamic presentation of two spatial sound stimuli, A and B. The total signifies the range of values for five subjects [5.6]. **(a)** Amplitudes, **(b)** Latency-differences

pressure levels were 70 dBA at a peak value. The interstimulus interval was 300 ms.

Dynamic differences due to the two spatial stimuli are clearly observed in amplitudes and latency-differences as shown in Fig. 5.4. It is remarkable that amplitudes of AEP under dynamic presentation of spatial stimuli were quite different from the amplitudes which were obtained under the fixed spatial condition (Fig. 5.2). This indicates that the right hemisphere may be aroused by dynamic spatial stimuli. The great difference in latencies obtained over the right hemisphere under the spatial stimuli shown in Fig. 5.4b may be interpreted as providing support for the hypothesis of asymmetry of hemisphere functioning to the spatial factor. The latency-differences from the right hemisphere were clearly increased as the IACC increased ($p < 0.01$, Signed-Ranks Tests).

It is worth noticing that several authors have reported that when speech signals with consonants were presented without any spatial variation, activities over the left hemisphere were greater than those over the right [5.7, 8]. These results suggest that complexity of sound signals in time domain causes dominant processing in the left hemisphere. The fact may also hold for continuous music signals with melodies [5.9], particularly when student musicians listen to such music, focusing on the sequential aspects of the melody [5.10].

These phenomena link the left hemisphere to complex sound signals and activities of the right hemisphere to spatial stimuli, and may relate to the subjective preference judgments in which the temporal and the spatial parameters affect preference independently, as discussed before.

5.2 Optimum Design Objectives

The optimum design objectives can be described in terms of the subjectively preferred sound qualities, which are related to the temporal and the spatial parameters describing the sound signals arriving at both ears. They clearly lead to the following comprehensive criteria for achieving the optimal design of concert halls.

5.2.1 Listening Level (Temporal-Monaural Criterion)

Since the listening level is expressed by the autocorrelation function of sound signals with the delay $\tau = 0$, it must be classified with the temporal criteria. The listening level is, of course, a primary criterion for listening to the sound fields in concert halls. The preferred listening level depends upon the music motif and the particular passage being performed. For example, the preferred levels obtained by 16 subjects are in peak ranges of 77 – 79 dBA for music motif A with a slow tempo, and 79 – 80 dBA for music motif B with a fast tempo. The most significant conclusion is, however, that preferred levels for all tempos and motifs do not greatly differ from about 79 dBA for peak values. In practice, of course, equal level distribution throughout the concert hall is recommended.

5.2.2 Early Reflections After Direct Sound (Temporal-Monaural Criterion)

In the investigations of sound fields with a single reflection, an approximate relationship was discovered between the autocorrelation function, the amplitude A of the reflection and the most preferred delay time $[\Delta t_1]_p$.

This relationship is closely approximated by the identity

$$[\Delta t_1]_p = \tau_p\,, \quad \text{such that} \tag{5.4}$$

$$|\phi_p(\tau)| \leqq kA^c\,, \quad \text{for} \quad \tau > \tau_p\,, \quad \text{or} \tag{5.5a}$$

$$|\phi_p(\tau)|_{\text{envelope}} \approx kA^c, \quad \text{at} \quad \tau = \tau_p \tag{5.5b}$$

where k = const (=0.1), c = const (=1.0), Fig. 5.5a and the envelope of autocorrelation function defined must decrease monotonously. This relationship also holds for sound fields with both multiple early reflections and with subsequent reverberations, like those in concert halls. It may be applied by finding a single reflection equivalent to the early reflections and the subsequent reverberation. Its pressure amplitude may be chosen by

$$A = \left(\sum_{n=1}^{\infty} A_n^2\right)^{1/2}, \tag{5.6}$$

and the delay time set equal to that of the strongest reflection, which is usually the first reflection, according to the inverse square law. Thus the preferred initial time delay is found to be identical with the time delay at which the envelope of the autocorrelation function (coherence of the direct sound)

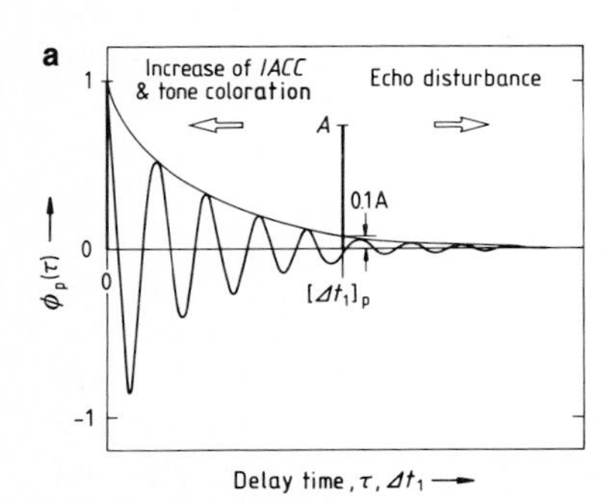

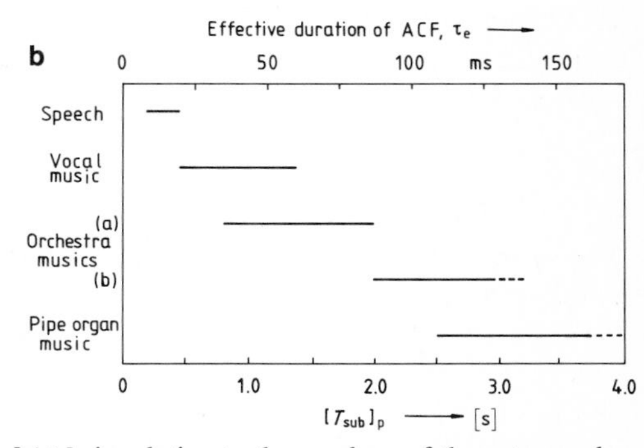

Fig. 5.5. (a) Preferred initial time-delay gap $[\Delta t_1]_p$ in relation to the envelope of the autocorrelation function. It is well described by the time delay τ_p such that $|\phi_p(\tau)| \leqq 0.1A$ for $\tau > \tau_p$. **(b)** Preferred reverberation times estimated for several sound sources

reaches a small value, $0.1A$ ($\tau_e = \tau_p$, when $A = 1$). If the envelope of the autocorrelation function is exponential, then (5.5) is simply expressed by

$$\tau_p \approx [1 - \log_{10} A]\,\tau_e\,. \tag{5.5c}$$

Also, the preferred delay of the second reflection, in relation to the direct sound, is approximately given by

$$[\Delta t_2]_p \approx 1.8\,\tau_p\,. \tag{5.7}$$

5.2.3 Subsequent Reverberation Time After Early Reflections (Temporal-Monaural Criterion)

For the flat frequency characteristics of reverberation, the preferred subsequent reverberation time is described by

$$[T_{\mathrm{sub}}]_p \approx 23\,\tau_e\,. \tag{5.8}$$

Initially, the coefficient in the equation was thought to depend on the total amplitude A. But, as mentioned in Sect. 4.3, it was found to be invariant in sound fields with two early reflections and subsequent reverberation.

The most preferred reverberation times estimated for each sound source are shown in Fig. 5.5b. A lecture and conference room must be designed for speech, and an opera house and similar theaters for vocal music. For orchestra music, these may be two or three types of concert-hall designs according to the effective duration of the autocorrelation function. For example, Symphony No. 41 by Mozart, "Le Sacré du Printemps" by Stravinsky and Arnold's Sinfonietta have short autocorrelation times and fit orchestra music of type (a). On the other hand, Symphony No. 4 by Brahms and Symphony No. 7 by Bruckner are typical of orchestra music (b). Much longer autocorrelation functions are typical for pipe organ music, for example, by Bach.

As is well known, earlier works on the optimum reverberation time, *Knudsen* [5.11] reported for speech signal maximizing its articulation. On the other hand, *Kuhl* [5.12] showed that the optimum reverberation time depended upon the type of music: 1.54 s for classical music (Symphony No. 41, "Jupiter" by Mozart), 2.07 s for romantic music (Symphony No. 4 by Brahms), and 1.48 s for modern music ("Le Sacré du Printemps" by Stravinsky). These results are probably best interpreted by referring to the effective duration of autocorrelation function of the source signals, as suggested by *Fourdouiv* [5.13]. Speech intelligibility in rooms has been predicted by the modulation transfer function [5.14] or the transmission envelope of any signal [5.15]. Considering the fact that the envelope of autocorrelation function is mainly determined by the envelope of source signals, the index with no noise conditions may probably be related to the autocorrelation function of the speech signal and music signals. The frequency characteristics of reverberation time seem to influence the multidimensional subjective space as reported by *Yamaguchi* [5.16]. The preference judgment may be performed in relation to the autocorrelation function of each frequency band.

5.2.4 Incoherence at Both Ears (Spatial-Binaural Criterion)

All available data indicates a negative correlation between the magnitude of the IACC and subjective preference. This holds only under the condition that the maximum value of the interaural cross correlation at $\tau = 0$ is maintained. If not, then image shift of the source will occur.

To obtain a small magnitude of the IACC in the most effective manner, the directions from which the early reflections arrive at the listeners should be kept within a certain range of angle from the median plane, i.e., $\pm(55° \pm 20°)$. It is obvious that the sound arriving from the median plane makes the IACC magnitude larger. Surprisingly, sound arriving from 90° in the horizontal plane is not always advantageous, because the similar "detour" paths around the head to both ears cannot decrease the interaural cross correlation effectively, particularly for frequency ranges higher than 500 Hz (see, Table D.3). Due to the need for a large interaural level difference and a proper interaural time difference, the optimum angles are, therefore, found in the range centered on 55° for commonly encountered music sources (Table D.1).

5.3 Theory of Subjective Preference

In general, when I-dimensional significant objective parameters, which are included in the sound signals at both ears in a concert hall, are given by $x_1, x_2, \ldots, x_I$ (Chap. 1), then the scale value of a one-dimensional subjective response is expressed by [5.17]

$$S = g(x_1, x_2, \ldots, x_I) \,. \tag{5.9}$$

For example, the scale value of subjective preference has been obtained by the law of comparative judgment. As explained in Chap. 4, each parameter contributes independently to the scale value of preference, therefore, (5.9) may be reduced to

$$S = \sum_{i=1}^{I} g(x_i) = \sum_{i=1}^{I} S_i \,, \tag{5.10}$$

where we set $S_i = g(x_i)$. This is analogous to the principle of superposition in a linear system, ex. (4.12). Without loss of generality, it may be conveniently assumed at the most preferred condition that $[g(x_i)]_p = 0$, $i = 1, 2, \ldots, I$ and thus $S_{\max} = 0$.

Before this theory is verified, there are four major problems to be solved:

1) To identify independent objective parameters of the I-dimensional preference function for the sound fields, which determine the subjective preference;

2) to find the most preferred conditions and to verify the independence of the parameters in the subjective preference space;
3) to find a consistent of the unit of scale values obtained from different test series using different music motifs; and
4) to find the functions $g(x_i)$, $i = 1, 2, \ldots, I$.

Items (1) and (2) have been discussed previously. We have identified four significant parameters which may be called the preference cue, i.e., $I = 4$, and their most preferred conditions. To obtain subjective preference at a design stage, we discuss Items (3) and (4) in this section.

The scale value of preference may be realized by normalizing each parameter using its most preferred value. In this way the data presented in Chap. 4 can be reduced to a general form [5.18]. All available data obtained with more than three plots in relation to each parameter will be employed here.

Scale values of preference as a function of listening level (S_1) are shown in Fig. 5.6. Obviously, similar behavior for the scale values of preference from the tests with different music motifs (music motifs A and B) is shown by normalizing the listening level by the most preferred conditions. The preference value may be formulated by

$$S_1 = g(x_1) \approx -\alpha_1 |x_1|^{3/2}, \tag{5.11}$$

where $x_1 = 20 \log P/[P]_p$, P being the sound pressure and $[P]_p$ signifying the most preferred sound pressure for each music source, and

$$\alpha_1 \approx \begin{cases} 0.07, & x_1 \geqq 0 \\ 0.04, & x_1 \leqq 0 \,. \end{cases} \tag{5.12}$$

Scale values of preference as a function of the initial time delay (S_2) are shown in Fig. 5.7. These values are obtained under the condition that the time delay of the second reflection is nearly satisfied by (5.7). But, it is not critical. In this figure, scale values plotted by different symbols, representing data from different test series, are reasonably consistent with each other showing consistency of the unit of scale values.

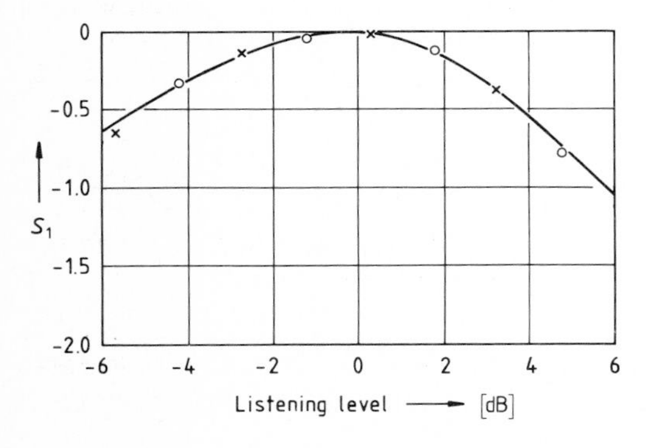

Fig. 5.6. Scale values of preference as a function of listening level. Data obtained by the test described in Sect. 4.3.2 [4.15]. (○) Music motif A; (×) music motif B. The smoothed curve is drawn through the plot and the scale value at the most preferred listening level is adjusted to zero

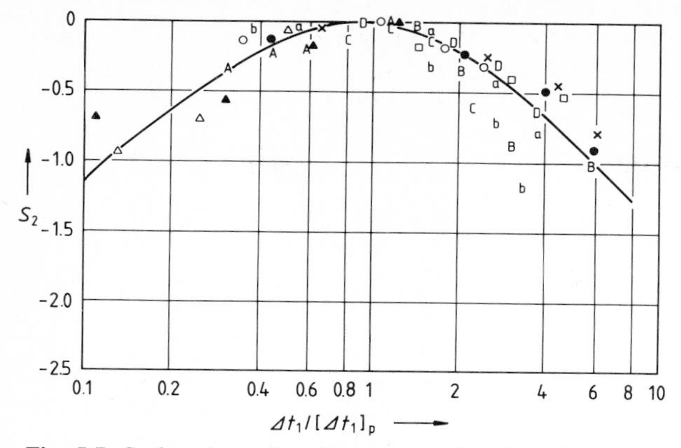

Fig. 5.7. Scale values of preference as a function of the initial time-delay gap Δt_1. Different symbols indicate the scale values obtained by different test series. Data (○, ×, □, ●) are obtained by the tests described in Sect. 4.3.1 [4.14]; (a, b) from [4.16]; (A, B, C, D) from [4.17]; and (△, ▲) from [4.13]. (○, a, A, △) Music motif A; (×, b, B, ▲) music motif B; (C, D) music motifs C and D; (□) music motif E; (●) speech S. The scale value at the most preferred delay time is adjusted to zero

The scale value of preference is also expressed in a similar form to (5.11)

$$S_2 = g(x_2) \approx -\alpha_2 |x_2|^{3/2}, \tag{5.13}$$

where $x_2 = \log \Delta t_1/[\Delta t_1]_p$, $[\Delta t_1]_p$ being given by (5.4 – 6) and

$$\alpha_2 \approx \begin{cases} 1.42, & x_2 \geqq 0 \\ 1.11, & x_2 \leqq 0 . \end{cases} \tag{5.14}$$

A slight preference may be found in the range of both x_1 and x_2 smaller than unity.

Scale values of preference as a function of the subsequent reverberation time (S_3) are shown in Fig. 5.8. No fundamental differences between the behavior of scale values of different sound sources are found. These results are obtained when $A = 4.1$. As indicated by the dashed curve in Fig. 5.9, when $A = 1.1$, the reverberation time becomes more critical in the range $T_{sub}/[T_{sub}]_p$ smaller than unity. But the most preferred condition is the same.

The scale values of preference may be found in the form of

$$S_3 = g(x_3) \approx -\alpha_3 |x_3|^{3/2}, \tag{5.15}$$

where $x_3 = \log T_{sub}/[T_{sub}]_p$, $[T_{sub}]_p$ being given by (5.7), and the coefficient α_3 may be expressed as a function of the total amplitude A, so that

$$\alpha_3 \approx \begin{cases} 0.45 + 0.74A, & x_3 \geqq 0 \\ 2.36 - 0.42A, & x_3 \leqq 0 . \end{cases} \tag{5.16}$$

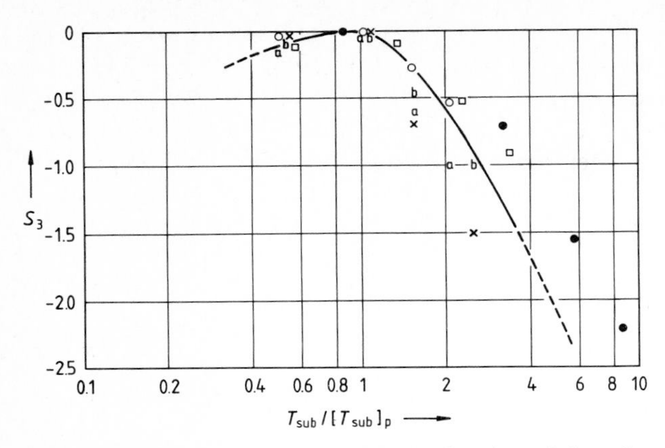

Fig. 5.8. Scale values of preference as a function of the subsequent reverberation time. Data (○, ×, □, ●) are obtained by the tests described in Sect. 4.3.1 [4.14]; and (a, b) from Sect. 4.3.3 [4.16]; (○, a) music motif A; (×, b) music motif B; (□) music motif E; (●) speech S. The scale values at the most preferred reverberation time are adjusted to zero

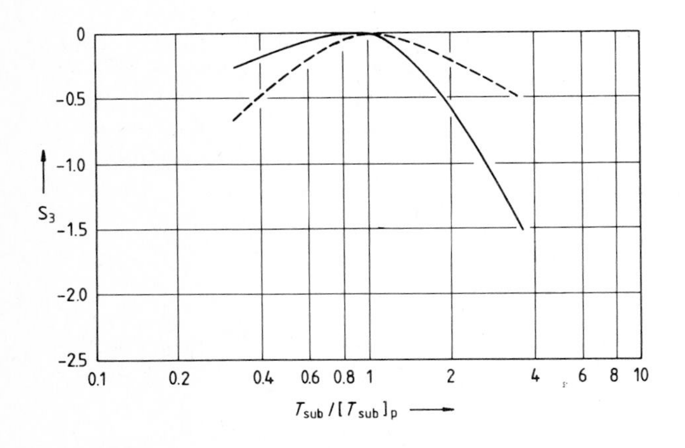

Fig. 5.9. Average scale values of preference as a parameter of the total amplitude of reflections A.
(———) $A = 4.1$;
(– – –) $A = 1.1$

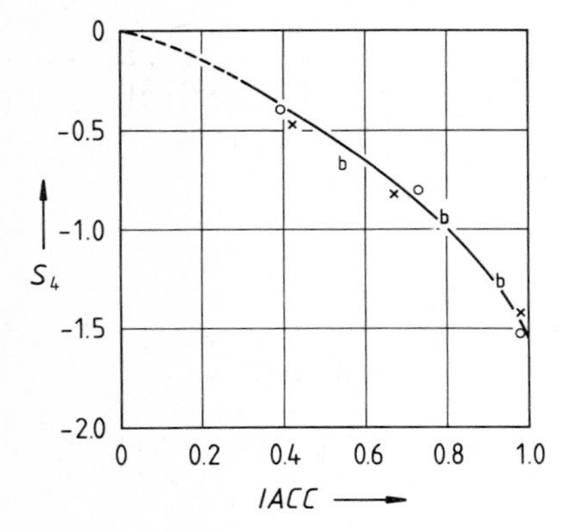

Fig. 5.10. Scale values of preference as a function of the IACC. Data (○, ×) are obtained by the tests described in Sect. 4.3.2 [4.15]; and data (b) from Sect. 4.3.3 [4.16]. (○) Music motif A; (×, b) music motif B. The maximum value of interaural cross correlation must be maintained at $\tau = 0$ to ensure frontal localization of the sound source

This holds for $A < 5$.

Scale values of preference as a function of the IACC (S_4) are shown in Fig. 5.10. Scale values, indicated by different symbols, are fairly close each other. Note that the values drop more rapidly when the IACC approaches unity. Hence, it is recommended to keep the IACC smaller than 0.5 for sound fields with a single source on the stage.

The scale value of preference is expressed by

$$S_4 = g(x_4) \approx -\alpha_4 (x_4)^{3/2}, \tag{5.17}$$

where $x_4 = \text{IACC}$ (3.8) and

$$\alpha_4 \approx 1.45 . \tag{5.18}$$

Applying variance analysis has established the statistical independence of the parameters between the delay of early reflections and subsequent reverberation time, the listening level and the magnitude of the IACC, and the reverberation time and the IACC [5.18]. Also, independence between the delay of early reflections and the IACC is found to be sufficient.

As for extreme music, reliabilities (95%) of scale values as plotted in Figs. 5.6 – 8, 10 are less than ± 0.16 (Music motif A) and ± 0.21 (Music motif B).

It is clear that the scale values obtained by different test series are resonably consistent with each other. This, in turn, reveals that the unit of scale values discussed in Sect. 4.1 is nearly constant. Therefore, the total scale value of preference may be obtained by

$$S \approx -\sum_{i=1}^{4} \alpha_i |x_i|^{3/2} . \tag{5.19}$$

We can now supply a negative value of preference discounted from the possible optimum condition at each seat prior to the final architectural design of concert halls.

Of particular interest is that the scale values of preference are all formulated in terms of the 3/2 power of the normalized parameters, expressed in the logarithm for the temporal-monaural criteria, x_1, x_2 and x_3. The spatial-binaural criterion x_4 is expressed by the term of 3/2 power of its *real* value.

Referring to the model of auditory pathways (Fig. 5.1), we can assume a further model of preference judgments as shown in Fig. 5.11. The first stage

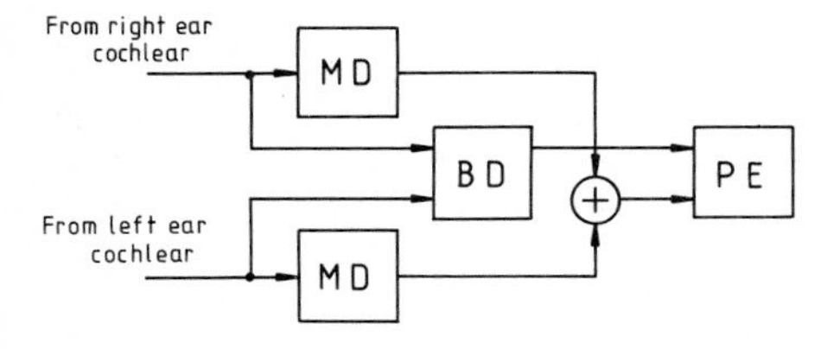

Fig. 5.11. A model of preference judgments. (MD) Monaural detector (logarithmic, x_1, x_2, x_3); (BD) binaural detector (linear, x_4); (PE) preference evaluator ($|x_i|^{3/2}$, $i = 1, 2, 3, 4$); ($\oplus$) signals are combined

of the model comprises a logarithmic-monaural detector based on the autocorrelation function process and a linear-binaural detector based on the cross-correlation process. The second stage of the model is an evaluator of preference which is postulated to be operative based on the "3/2 power process" somewhere in the upper part of the brain. The temporal monaural detector corresponds to Weber-Fechner's law, but the binaural detector is linearly operative and has a more critical effect on subjective judgments.

To obtain *more* preferred conditions, it is unlikely that there are other significant and independent objective parameters on the subjective preference space, even though other parameters less pronounced than the four have been found in existing concert halls (Sect. 7.4). However, if an additional objective parameter which affects the preference are found, such as, for example, spectral effects of reverberation, then its scale value, obtained in a manner similar to that with two variable parameters (including one used in the previous chapter), may be added to the total scale value of preference.

5.4 Calculating Subjective Preference for a Concert Hall

As a typical example, we shall discuss the quality of the sound field in a concert hall with a shape similar to the Symphony Hall in Boston. The plan of the hall is outlined in Figs. 5.12 – 20. The hall is 25 m wide, 50 m long and 18 m high. For simplicity, reflections and scattering by balconies and floor are not considered. Also, it is supposed that a single source is located at the center, 1.2 m above the stage floor, which is 1 m above the audience floor level. Receiving points at a height of 1.1 m above the floor level correspond to the ears' positions.

According to the image method, thirty reflections with their amplitudes, delay times, and directions of arrival at the listeners were taken into account. The sound pressure level SPL is given by

$$\begin{aligned} \mathrm{SPL} &= \mathrm{PWL} + \mathrm{L}\,, \\ \mathrm{L} &= 10\log(1+A^2) - 20\log d_0 - 11\ [\mathrm{dB}]\,, \end{aligned} \tag{5.20}$$

where PWL is the power level of the source [$= 10\log(W/10^{-12})$, W being the acoustic power of the sound source]. Here A is the total amplitude of reflections, see (5.6), and $d_0(=|r-r_0|)$ is the distance between the source and the listeners' positions. Since the power level itself is unknown at the design stage of concert halls, only the term L can be calculated for each seat. Contour lines of equal L values are shown in Fig. 5.12, and those of equal total amplitude A in Fig. 5.13. The rapid attenuation near the source for $d_0 < 12$ m is due to the inverse square law. The value of L is almost independent of the distance and it is the range of ± 1.5 dB, $d_0 \geqq 12$ m.

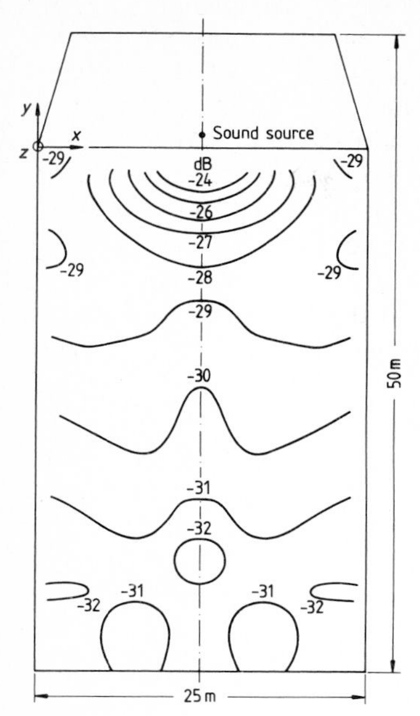

Fig. 5.12

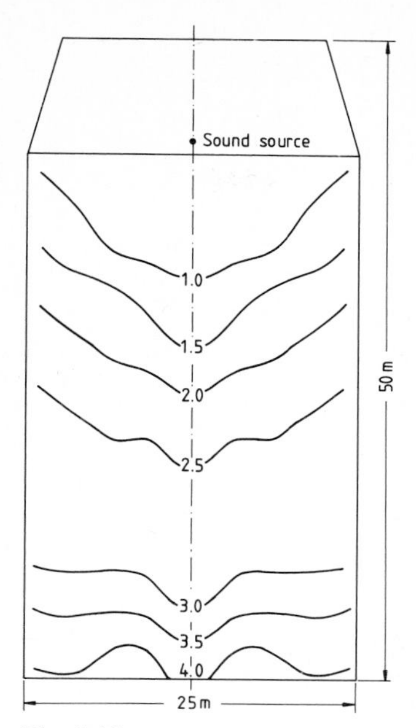

Fig. 5.13

Fig. 5.12. Contour lines of equal listening level L calculated with (5.20)

Fig. 5.13. Contour lines of equal total amplitude of reflections A (5.6)

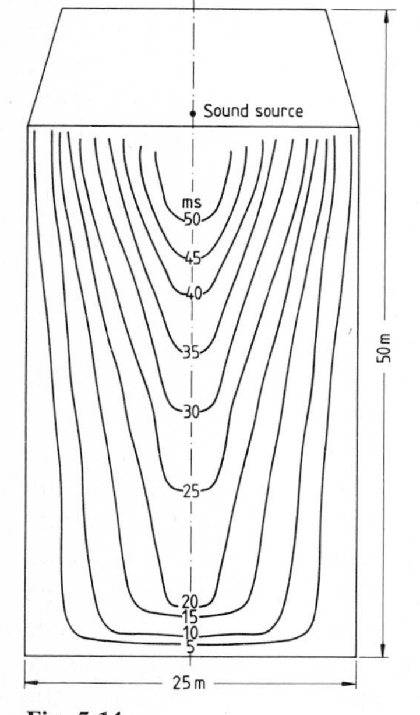

Fig. 5.14

Fig. 5.14. Contour lines of equal initial time-delay gap Δt_1 calculated with (5.21)

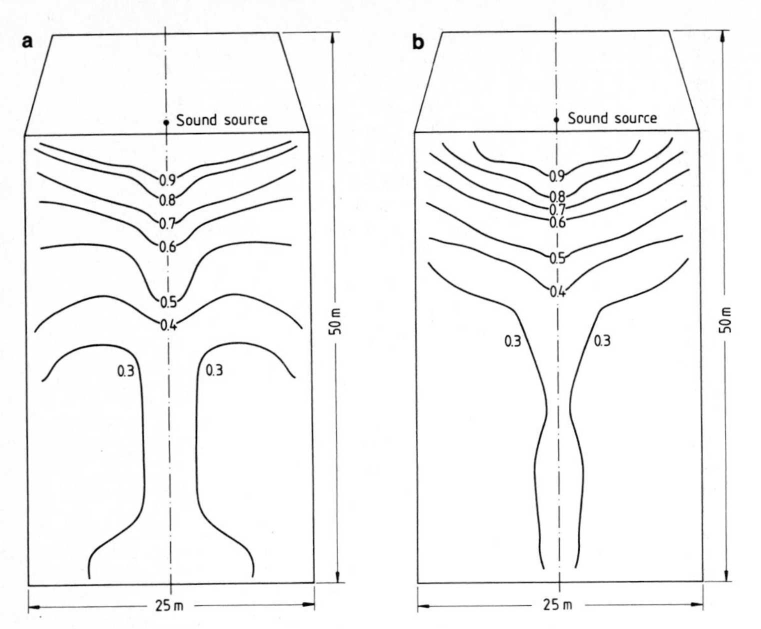

Fig. 5.15 a, b. Contour lines of equal IACC calculated with (3.7, 8) with correlation values listed in Table D.1. (**a**) Music motif A; (**b**) music motif B

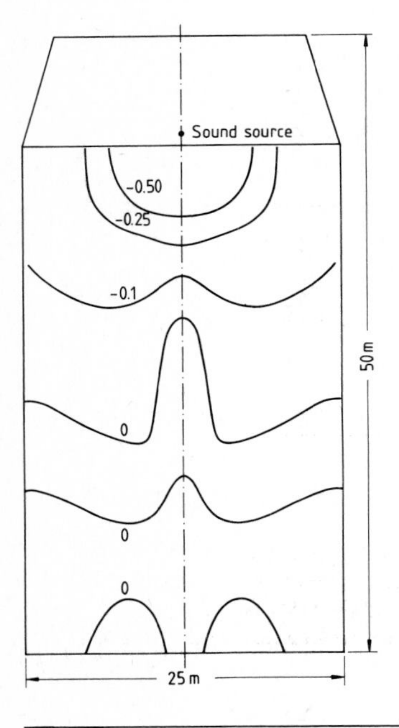

Fig. 5.16. Contour lines of equal subjective preference for the relative listening level calculated with (5.11, 12). The preferred level is assumed at $L = -30$ dB (Fig. 5.12)

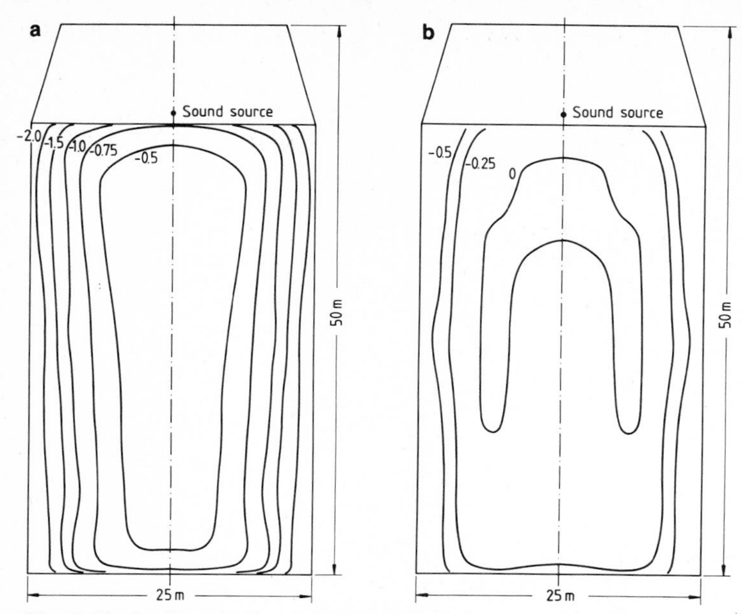

Fig. 5.17 a, b. Contour lines of equal preference for the initial time-delay gap calculated with (5.13, 14). The area tending to zero is preferred. Note that the total amplitude of reflection given by (5.6) increases with increasing distance between the source point and the listener. Therefore shorter initial time-delay gaps than those in Fig. 5.14 for music motif B are preferred at the rear seats. **(a)** Music motif A; **(b)** music motif B

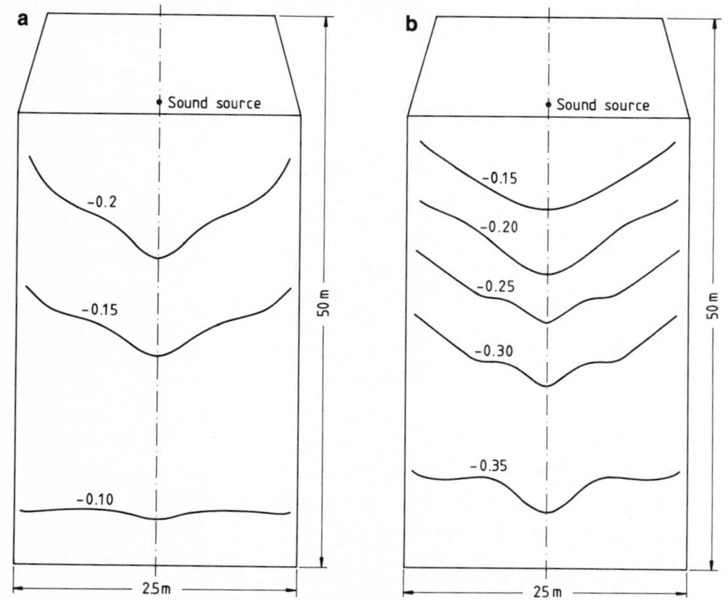

Fig. 5.18 a, b. Contour lines of equal preference for the subsequent reverberation time calculated with (5.15, 16). **(a)** Music motif A; **(b)** music motif B

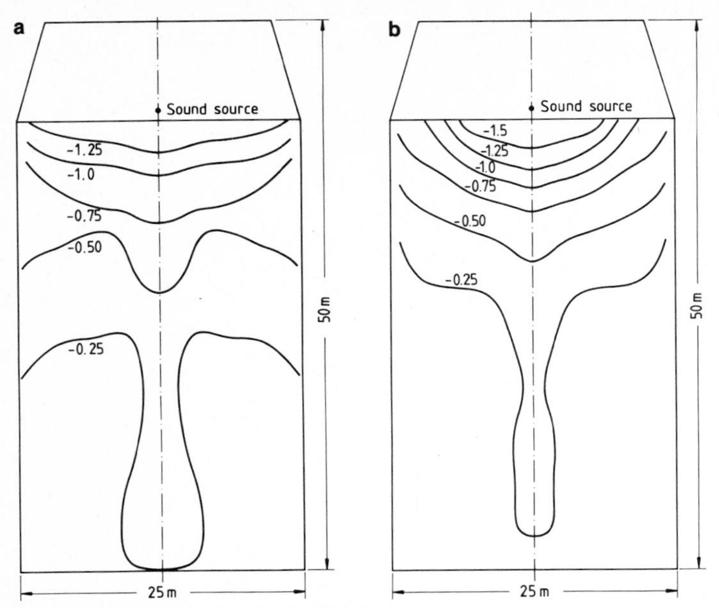

Fig. 5.19 a, b. Contour lines of equal preference for the IACC calculated with (5.17, 18). **(a)** Music motif A; **(b)** music motif B

The initial time-delay gap between the direct sound and the first reflection is in simply obtained by

$$\Delta t_1 = (d_1 - d_0)/c\,, \tag{5.21}$$

where d_1 is the path length of the first reflection. Contour lines of equal time delay are shown in Fig. 5.14. It is quite natural that the largest time delay occurs on the center line near the source.

The IACC at each seat is calculated via (3.7, 8). The values of correlations needed for the calculation are listed in Table D.1. To find correlation values for each single sound reflection arriving from any direction (ξ, η) at the listener, the angle measured from the median plane given by $\sin^{-1}(\sin\xi\cos\eta)$ is substituted for the horizontal angle ξ. This approximation is reasonable, because the interaural cross correlation is mainly dependent upon the angle from the median plane.

Contour lines of the equal IACC value are shown in Fig. 5.15. The values on the center line are larger than in other areas. This tendency is probably emphasized in existing concert halls because coherent reflections arrive simultaneously at listeners seated near the center line.

Utilizing the functions described by the smooth curves in Figs. 5.6 – 10 or (5.11 – 18), we can obtain a scale value of subjective preference for each objective parameter at each seat.

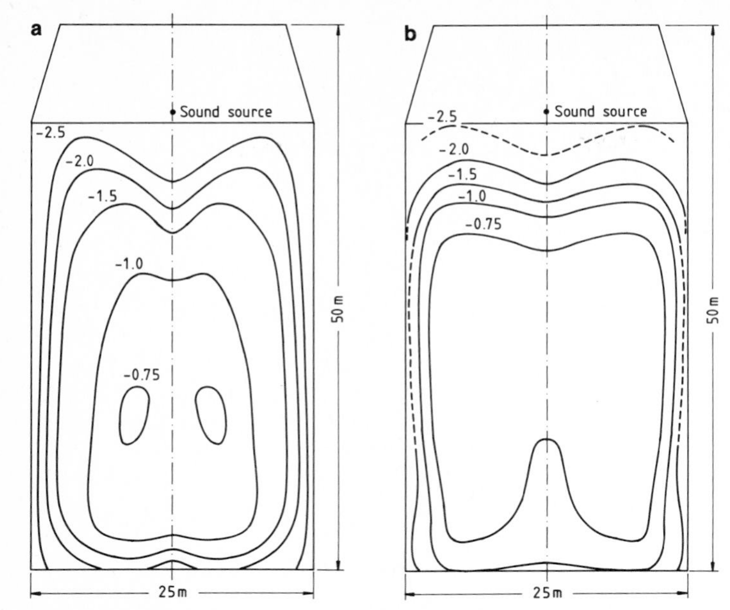

Fig. 5.20a, b. Contour lines of total subjective preference for the four physical parameters calculated with (5.19). (**a**) Music motif A; (**b**) music motif B. The area with total subjective preference greater than -1.0 for music motif A is smaller than that for music motif B, because of the effective duration of autocorrelation function

Assuming that the preferred listening level is obtained on the center line in the hall 20 m from the source position (at $L = -30$ dB in Fig. 5.12), the scale values of preference at all other seats may then be calculated. Results are illustrated in Fig. 5.16. This procedure may be applied for any kind of music source because the distribution of the listening level is essentially uniform for different sound sources.

After determining the most preferred initial time delay $[\Delta t_1]_p$ by (5.4 – 6), the scale value of preference at each seat may be calculated by (5.13, 14). Contour lines of equal preference for music motifs A and B are shown in Fig. 5.17. For music motif A, there is no optimum condition in this parameter, because of the long-duration autocorrelation function. On the other hand, the effective duration of the autocorrelation function of music motif B is short enough for a hall with a width of 25 m so that an optimum area with zero score does exist but is limited in a tuning-fork shape as shown in Fig. 5.17b. Note that the total amplitude of reflections A shown in Fig. 5.13 rises with increasing distance from the source position, and that this causes the decline of delay-time preference in the rear seats for music motif B.

The scale values of preference for the subsequent reverberation time is calculated by (5.15, 16) together with the total amplitude A. The resulting preference distribution for the two motifs is shown in Fig. 5.18. Usually the reverberation time in halls is almost constant, and the corresponding scale values of preference do not vary much with seat position. Slight variations are due to A in (5.16). In this calculation, the actual reverberation time is assumed to be 1.8 s throughout the hall.

The preferred reverberation time $[T_{sub}]_p$ is 2.9 s for music motif A and 0.99 s for music motif B, thus values of $T_{sub}/[T_{sub}]_p$ are 1.6 and 0.5, respectively.

The scale values of preference for the IACC at each seat is obtained from (5.17, 18). The contour lines of equal preference in the hall for the two motifs are shown in Fig. 5.19. A similar spatial distribution in the hall can be seen for both motifs because the binaural criterion does not vary with music motif.

Based on the principle of superposition given by (5.19), the total scale value of preference for the sound fields in the hall can be calculated. Therefore, we can predict subjective preference at the design stage of halls.

Contour lines of the total scale value of preference for the two motifs in the hall are shown in Fig. 5.20. Obviously, as different music is performed in a given concert hall, so the total preference value changes according to the autocorrelation function of the music. Comparing the results for music motifs A and B, motif B comes to better advantage in this concert hall, but neither is precisely suited to the size and shape of this hall. The most preferable and suitable music may be selected by means of the effective duration of the autocorrelation function. For example, music motif D (Symphony No. 102 by Haydn) with $\tau_e = 65$ ms would seem to be one of the most suitable pieces of music to be performed at Boston Symphony Hall.

So far we have discussed the subjective preference for a single sound source on the stage, but we should note that the scale values of preference may be calculated for every sound source location on the stage.

6. Design Study

Fundamental design studies based on the four design objectives will now be described with examples, along with some guidelines for acoustic design in concert halls.

6.1 Walls and Ceiling

The side walls are the most important surfaces in supplying useful early reflections to listeners in concert hall. To minimize the IACC, the side walls must be essentially reflective, and the directions of the reflected sound to the listeners should be adjusted and centered on $\pm 55°$. Furthermore, to keep the frequency spectrum of the reflected sound similar to that of the direct sound, certain minimum dimensions for wall surfaces are required (Fig. 2.4, Sect. 2.2).

An example for improving the reflected sound of a wall is shown in Fig. 6.1. The wall consists of slanted reflective surfaces with the maximum distance from the original wall at the upper-rear edge [6.1].

If the walls are symmetrical and if the sound source is located at the center of the stage, then the IACC for the listeners sitting on or near the center line is

Fig. 6.1. Improvement of early reflection from side wall in the Mimasaka-Bunka Center [6.1]. The stage is on the right-hand side in the figure

increased because identical sound signals arrive at both ears simultaneously. This can be rectified by carefully choosing the positions of the sound sources on the stage, the distribution of acoustic material, as well as the shape of the walls. As far as binaural hearing is concerned, therefore, the best course is to design acoustically asymmetrical properties.

On the other hand, strong reflections from the ceiling and rear wall increase the IACC. But simply to absorb them is inadequate, because we also need a sufficiently generous sound energy supply for the seats in the rear half of the hall. For this purpose, *Schroeder* [6.2] proposed a highly diffusing ceiling to reflect most of the sound energy to the side walls, so that it would arrive at the listeners from a suitable direction for minimizing the IACC. The theory is based on the pseudo-random sequences, for example, the quadratic-residue sequences of an elementary number theory originally investigated by Legendre and Gauss.

Returning to Sect. 2.3, where reflection from periodic wall structure is treated, the velocity potential of a reflected wave at $y=0$ is reduced from (2.26) to

$$U_{\mathrm{r}}(x;\omega)=\sum_{r=-\infty}^{\infty} R_r \exp(\mathrm{j}\,\alpha_r x)\,, \tag{6.1}$$

where $\alpha_r=\alpha_0+2\pi r/l$, $\alpha_0=k\sin\theta$, $\phi=\pi/2$, and the condition of undamped oscillations implies $-(1+\sin\theta)\,l/\lambda\leqq r\leqq(1-\sin\theta)\,l/\lambda$. Equation (6.1) may be rewritten as

$$\begin{aligned}U_{\mathrm{r}}(x;\omega)&=\exp(\mathrm{j}\,\alpha_0 x)\sum_{r=-\infty}^{\infty} R_r \exp\left(\frac{\mathrm{j}2\pi r x}{l}\right)\\&=U_{\mathrm{i}}(x;\omega)\,U_{\mathrm{s}}(x;\omega)\,,\end{aligned} \tag{6.2}$$

where $U_{\mathrm{i}}(x,\omega)=\exp(\mathrm{j}\,\alpha_0 x)$ is the incident wave and

$$U_{\mathrm{s}}(x;\omega)=\sum_{r=-\infty}^{\infty} R_r \exp\left(\frac{\mathrm{j}2\pi r x}{l}\right). \tag{6.3}$$

Thus, $U_{\mathrm{s}}(x;\omega)$ is regarded as a reflection factor. Since $U_{\mathrm{s}}(x;\omega)$ is periodic with respect to x, we therefore have the Fourier series

$$R_r=\frac{1}{l}\int_0^l U_{\mathrm{s}}\exp\left(\frac{-\mathrm{j}2\pi r x}{l}\right)dx\,. \tag{6.4}$$

Now let us consider a periodic cross section with a period N as shown in Fig. 6.2, $w(\ll\lambda)$ being a width of each "well". If we assume a locally reacting surface, then (6.4) at $\theta=0°$ becomes [6.2]

$$R_r=\frac{1}{Nw}\int_0^{Nw}\frac{Z(x;\omega)-\varrho c}{Z(x;\omega)+\varrho c}\exp\left(\frac{-\mathrm{j}2\pi r x}{Nw}\right)dx\,, \tag{6.5}$$

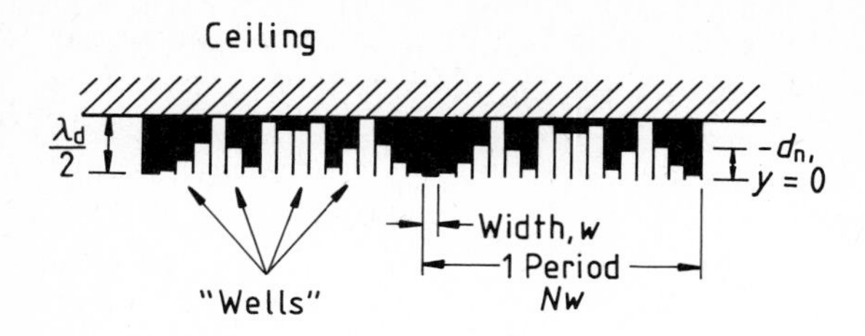

Fig. 6.2. Quadratic residue diffuser based on quadratic residues of $N=17$, for wideband broad sound diffusion from ceiling [6.2]

where we set

$$U_s(x;\omega)=\frac{Z(x;\omega)-\varrho c}{Z(x;\omega)+\varrho c}.$$

The impedance at each well is expressed by

$$Z(x;\omega)=-\mathrm{j}\varrho c/\tan[2\pi d(x)/\lambda],$$

where $d(x)$ is the depth of each well. After algebraic manipulation we obtain

$$R_r=\frac{1}{Nw}\int_0^{Nw}\exp\left(\frac{-\mathrm{j}4\pi d(x)}{\lambda}-\frac{\mathrm{j}2\pi rx}{Nw}\right)dx. \tag{6.6}$$

The problem is to determine the depth $d(x)$ for high diffusion. A good choice is based on the quadratic-residue sequence for $d(x)$, defined as

$$d_n=(\lambda/2N)s_n, \tag{6.7}$$

where $s_n=n^2$ (n^2 is taken as the least nonnegative remainder modulo N, and N is an odd prime).

For example, for $N=17$, starting with $n=0$, the sequence reads

0, 1, 4, 9, 16, 8, 2, 15, 13, 13, 15, 2, 8, 16, 9, 4, 1 ,

so that the period is N.

These sequences have the following property. The discrete Fourier transform of the exponential sequence

$$c_n=\exp(\pm\mathrm{j}2\pi s_n/N) \quad \text{is expressed by} \tag{6.8}$$

$$|C_m|^2=\left|\frac{1}{N}\sum_{n=1}^{N}c_n\exp\left(\frac{-\mathrm{j}2\pi nm}{N}\right)\right|^2=\frac{1}{N}. \tag{6.9}$$

If we call the depth of the nth well $-d_n$ instead of $d(x)$, then the Fourier integral of (6.6) can be approximated by a sum as given by (6.9), thus

$$|R_r|^2 = \text{const} = \frac{1}{N} . \tag{6.10}$$

Under the condition of (6.10), the scattering amplitude of the rth spectral order is the same. Therefore, the quadratic residue surface is regarded as an optimum diffuser. Note that for oblique incidence (6.7) may be replaced by [6.3]

$$\begin{aligned} d_n &= \left(\frac{\lambda}{2\pi}\right) \tan^{-1}\left(\cos\theta \tan\left[\frac{\pi}{N} n^2 \operatorname{mod}\left\{\frac{2N+1}{N}\right\}\right]\right) \\ &= \left(\frac{\lambda}{2\pi}\right) \tan^{-1}\left[\cos\theta \tan\left(\frac{\pi s_n}{N}\right)\right] . \end{aligned}$$

The transfer function for the reflection from the surface may be analyzed by the theory treated in Sect. 2.4. Calculations for the scattering characteristics of Schroeder's diffuser have been carried out by *Strube* [6.4, 5]. Some examples of calculated values and measured curves are shown in Fig. 6.3.

The design procedure of the optimum diffuser at $\theta = 0°$ is as follows.

1) First decide the diffusing frequency range f_{low} and f_{high}, $f_{\text{low}} < f_{\text{high}}$; then the minimum period N is approximately given by $f_{\text{high}}/f_{\text{low}}$. Also, the width w of each well should not exceed $c/2f_{\text{high}}$.

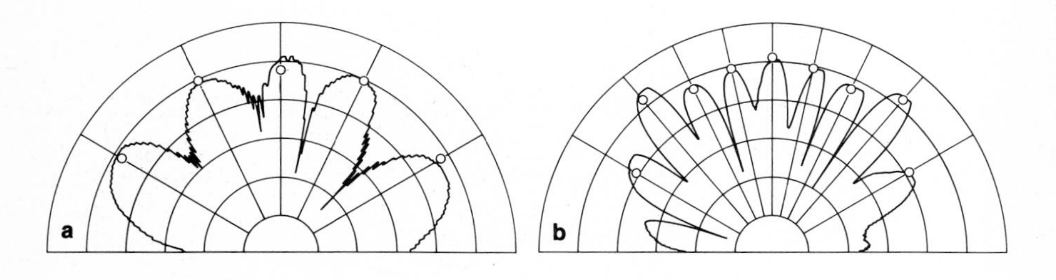

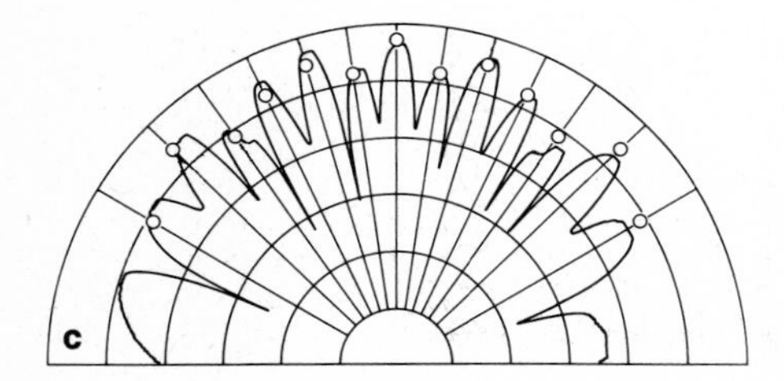

Fig. 6.3a–c. Measured and calculated reflection pattern of the scale model of quadratic residue diffuser, $N = 17$, $\lambda_d = 3$ cm, $\theta = 0°$, $\phi = 0°$ [6.4, 5]. **(a)** $\lambda_d/\lambda = 1.0$; **(b)** $\lambda_d/\lambda = 2.0$; **(c)** $\lambda_d/\lambda = 3.0$

2) From (6.7), the depth of each well d_n is given by

$$d_n = \frac{\lambda_d}{2N} s_n , \tag{6.11}$$

where λ_d is called the design wavelength ($= c/f_{low}$).

For example, if we choose a frequency range such that $f_{low} = 200$ Hz ($\lambda_d = 170$ cm) and $f_{high} = 2200$ Hz, then $N = 11$ and $w = 7.7$ cm. The depth of each well should be as follows:

$$d_n = 7.7\, s_n\, [\mathrm{cm}]$$
$$= 0, 7.7, 30.9, 69.5, 38.6, 23.2, 23.2, 38.6, 69.5, 30.9, 7.7\, [\mathrm{cm}] ,$$
$$n = 0, 1, 2, \ldots, 11 ,$$

where

$$s_n = 0, 1, 4, 9, 5, 3, 3, 5, 9, 4, 1 .$$

The length of period l becomes $Nw = 85$ cm.

6.2 Floor and Seats

6.2.1 Sound Transmission Over Seat Rows

According to the theoretical analysis described in Sect. 2.3, the sound transmission characteristics over seat rows can be calculated and modifications can

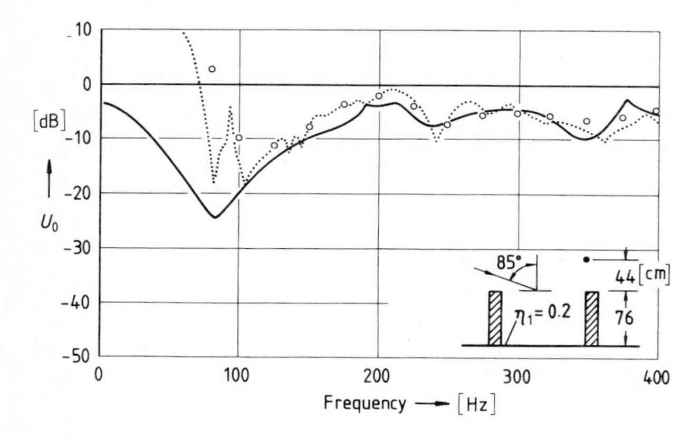

Fig. 6.4. Comparison between measured and calculated sound pressures, with reference to the sound pressure of direct sound [6.6]. (○ ○ ○) Tone-burst measurements with a 1/10 scale model [6.7]; (·········) continuous-wave measurements with the 1/10 scale model [6.7]; (———) calculations with a similar boundary condition in the measurement ($l = 82$ cm, $d_1 = 76$ cm, $h_1 = 76$ cm)

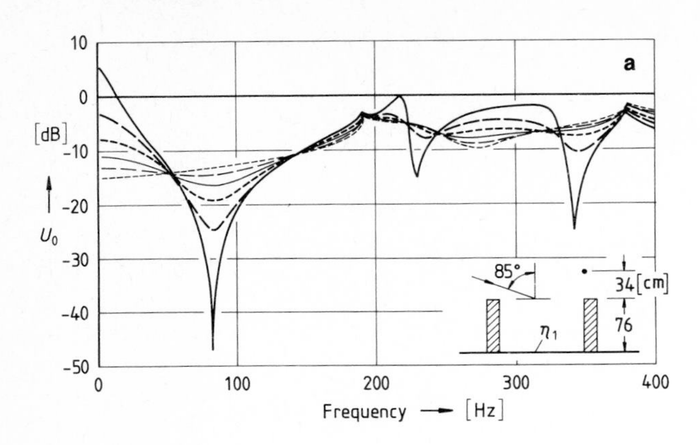

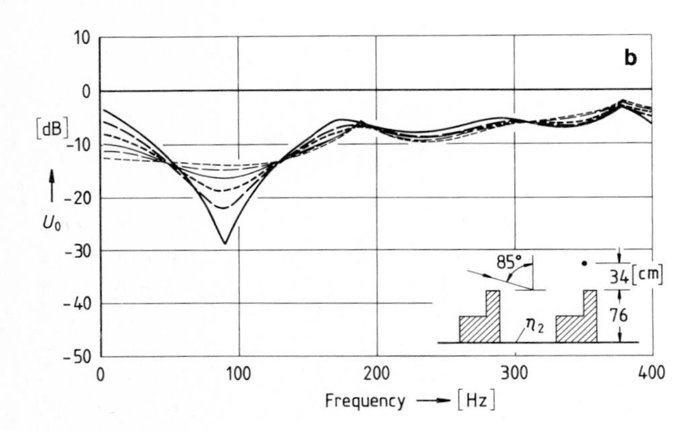

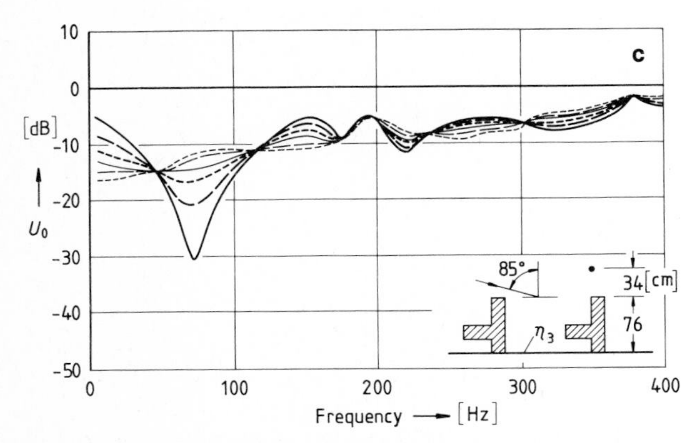

Fig. 6.5 a – c. Calculated sound pressures over a seat row with the specific acoustic admittance of the floor surface as a parameter [6.6]. (———) 0.01; (– – –) 0.2; (-----) 0.4; (———) 0.6; (– – –) 0.8; (---------) 1.0. **(a)** Seat section (a); **(b)** seat section (b); **(c)** seat section (c). Seat sections are shown in the lower part of each figure (l = 90 cm)

be evaluated which will reduce the low-frequency attenuation [6.6]. In the calculation, the horizontal angle of incidence ϕ is fixed at 90°.

To verify the theoretical model the calculated values are compared with measured values obtained from a 1/10 scale model at Bell Laboratories [6.7]. The measured and calculated relative sound pressure levels above the seats with no underpass (referred to as the sound pressure of the direct sound) are compared in Fig. 6.4 as a function of frequency. For the measured values in the figure, the frequency is multiplied by a factor of 0.1. The specific acoustic admittance of the floor surface η_1 is set at 0.2 in this calculation. Note that the sound pressure levels calculated are almost independent of the acoustic admittance for the frequencies above 100 Hz, except for $\eta_1 \to 0$ in Fig. 6.5a. Measured and calculated values in Fig. 6.4 agree closely except for the frequencies below 100 Hz where the sound pressures measured using the scale model increase with respect to those calculated. This increase is in part caused by a diffraction effect of the finite floor dimension of the model tested.

6.2.2 Effects of Seat Configuration and Floor Absorption

The two-dimensional configurations of the seat rows are shown in the lower part of Fig. 6.5a – c. Sections (a), (b) and (c) consist of one, two and three rectangular regions in $y < 0$, respectively. Section (a) corresponds to the seat bottoms raised and Sections (b) and (c) correspond to those opened but without any underpass. Also, calculated sound pressure levels for the three sections at $\theta = 85°$ are correspondingly shown in the figures. The levels are plotted with the specific acoustic admittance of the floor surface as a parameter, and show the effects of surface absorption. Similar tendencies are observed for all seat configurations. The maximum attenuations in the frequency range 70 – 90 Hz diminish with increasing absorption by the floor, while there are no fundamental differences in the attenuation tendencies at low frequencies.

6.2.3 Effects of the Angle of Wave Incidence

The vertical angle of wave incidence θ is often close to grazing incidence. The vertical angle was varied in the range of 70° – 89° in the calculations. As a rough estimate of the absorption of real floors and seats, the acoustic admittance of the floor was fixed at 0.2. The results are shown in Fig. 6.6.

The sound pressure throughout the calculated frequency range decreases uniformly with increasing angle of incidence. This phenomenon is called the Békésy effect [6.8]. When the angle is kept smaller than 80°, the excess attenuation can be less than 4 dB except around the dip frequency.

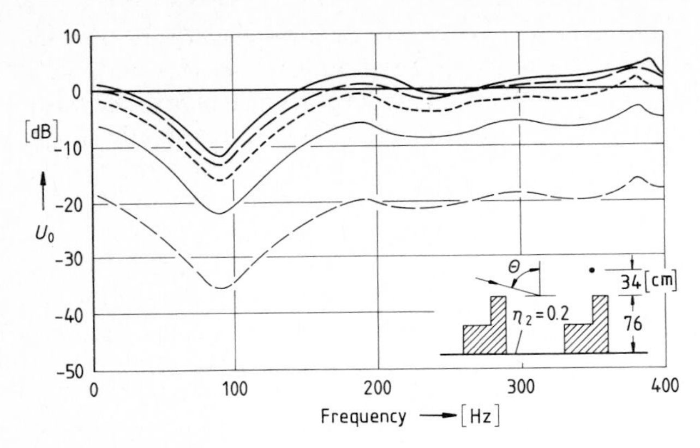

Fig. 6.6. Calculated sound pressures over the seat row with the angle of incidence θ as a parameter. The specific acoustic admittance of the floor surface is fixed at 0.2 ($l = 90$ cm) [6.6]. (———) 70°; (– – –) 75°; (-----) 80°; (———) 85°; (– – –) 89°

6.2.4 Effects of a Slit Resonator Under the Floor

In the previous section, it was shown that the large attenuation which occurs at the frequencies 80 ± 10 Hz can be reduced by making the floor surface absorptive. To realize this in the low-frequency range, effects of slit resonators installed under the floor were examined.

Figure 6.7 shows the sound pressure, at ear level without seats, as a parameter of the angle of incidence. There are no significant dips, indicating that absorption by the floor with a slit resonator is sufficient to remove the transmission dip. Such characteristics are also demonstrated in calculations *with* seats.

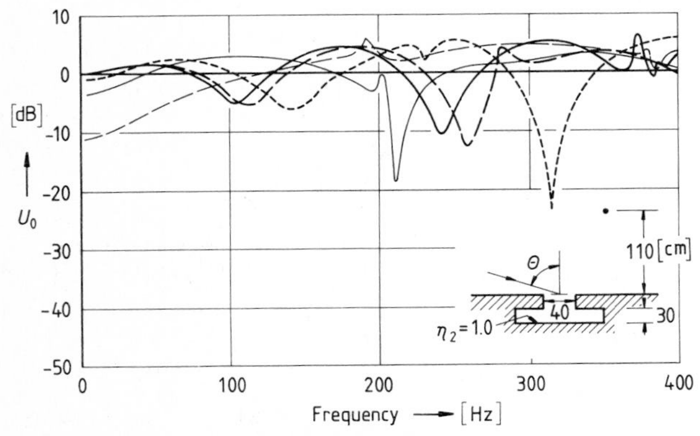

Fig. 6.7. Calculated sound pressures over a slit resonator under the floor with the angle of incidence θ as a parameter (without any seats) ($l = 90$ cm) [6.6]. (———) 1°; (– – –) 21°; (------) 41°; (———) 61°; (– – –) 81°. No significant dips are observed here

A possible alternative for improving the sound transmission over the seats would be to replace the floor by a grating, making the floor acoustically free. Leaving an air gap, the bottom then has diffusing characteristics similar to those of the ceiling and rear wall. If the space below the ears is designed similarly to the upper space, because of the equal importance of both spaces, then a fantasy may be realized!

6.3 Stage Enclosure

The other important problem in concert hall design is the acoustical environment of the musicians during the performance. The problem is mainly related to the stage design. The effect of stage size on the ensemble playing was investigated by *Marshall* et al. [6.9]. The stage sizes simulated and the preference results (ease of ensemble) are shown in Fig. 6.8a, b. It is clear that an optimum stage size for performers exists, so that configuration B is the most preferred condition.

The preferred time delay for the stage-size study, as in the auditorium studies, was assumed to be related to the autocorrelation function of the music signal being performed. *Nakayama* [6.10] showed that a similar concept in the temporal criteria to the listeners may be applied. The most preferred time delay of a single reflection for alto-recorder soloists was obtained by

$$[\Delta t_1']_p = \tau_p' , \qquad \text{such that} \tag{6.12}$$

$$|\phi_p(\tau)|_{\text{envelope}} \approx p A_1'^{\,q}, \quad \text{at} \quad \tau = \tau_p' \tag{6.13}$$

where $p = \text{const}$ ($= 0.3$), $q = \text{const}$ ($= 1.0$) and A_1' is the pressure amplitude of reflection in which the reference $A_1' = 1$ ($= 0$ dB) is defined by the aurally equal amplitude to that the direct sound.

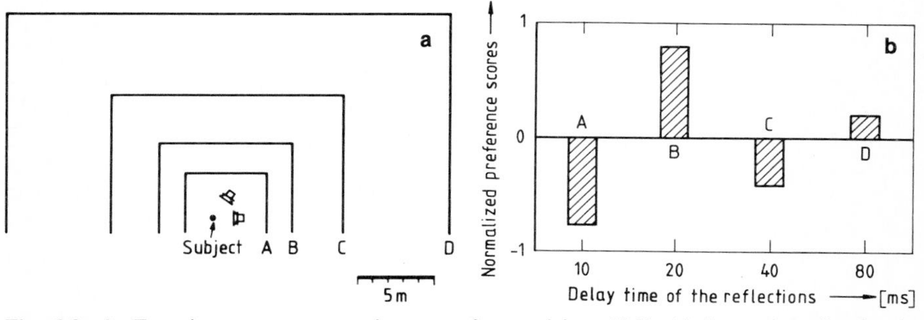

Fig. 6.8a, b. Experiment on stage environment for musicians [6.9]. (**a**) Stage sizes simulated; (**b**) preference scores. A, B, C, and D signify stage sizes, and amplitudes of reflection are −10 dB, −14 dB, −18 dB and −24 dB, respectively

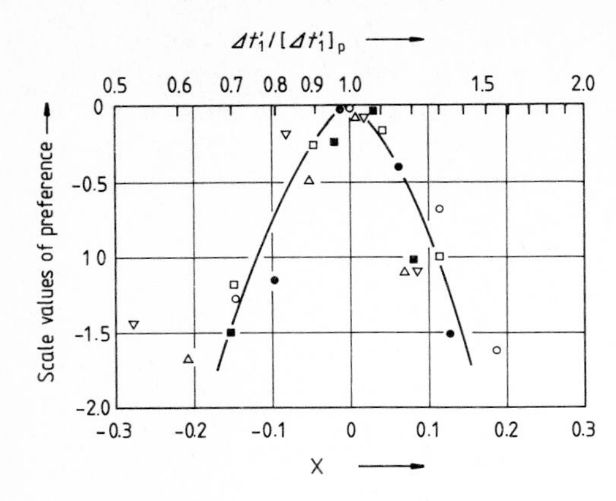

Fig. 6.9. Scale values of preference in ease of playing alto-recorder as a function of the time delay of single reflection $\Delta t_1'$ [6.10]. Different symbols indicate the scale values obtained by different test series. Results with music motif A′ with $\tau_{0.3} \approx 35$ ms, in which $\tau_{0.3}$ is defined by the delay at which the envelope of the normalized autocorrelation function becomes 0.3: (○) $A_1' = 0$ dB; (●) $A_1' = -6$ dB; (△) $A_1' = -12$ dB. Results with music motif B′ with $\tau_{0.3} \approx 50$ ms: (□) $A_1' = 0$ dB; (■) $A_1' = -6$ dB; (▽) $A_1' = -12$ dB

If A_1' is replaced by A_1 defined by the pressure amplitude which is determined by the $(1/r)$-law as in Chap. 3, then (6.13) is approximately reduced by

$$|\Phi_p(\tau)|_{\text{envelope}} \approx A_1, \quad \text{at} \quad \tau = \tau_p. \tag{6.14}$$

Normalizing the delay time by its most preferred delay time, the scale values of preference are shown in Fig. 6.9. The scale values obtained by different series of preference tests with different music pieces are surprisingly consistent with each other. Thus, an approximate formula representing the scale values for a performer may be expressed well by

$$S' \approx -\alpha' |x|^{3/2}, \tag{6.15}$$

where $x = \log \Delta t_1'/[\Delta t_1']_{p'}$ and

$$\alpha' \approx \begin{cases} 27.5, & x \geqq 0 \\ 25.0, & x < 0. \end{cases} \tag{6.16}$$

These values of α' are much greater than those for the audience listener's condition with the single reflection. The approximate values of α' for listeners are

$$\alpha' \approx \begin{cases} 2.2, & x \geqq 0 \\ 1.5, & x < 0. \end{cases} \tag{6.17}$$

Therefore, it is considered that the performers are more sensitive to the reflection than the listeners. Of particular interest is that the performer's preference seems to be increased with increasing IACC, in contrast with the results of listeners [6.11]. This shows that reflections from the ceiling and the back wall of the stage enclosure are much more important for performers than those from side reflectors on the stage. And, the most preferred delay time according to the effective duration of the autocorrelation function may be successfully adjusted by the height of the stage ceiling and/or the depth of the stage.

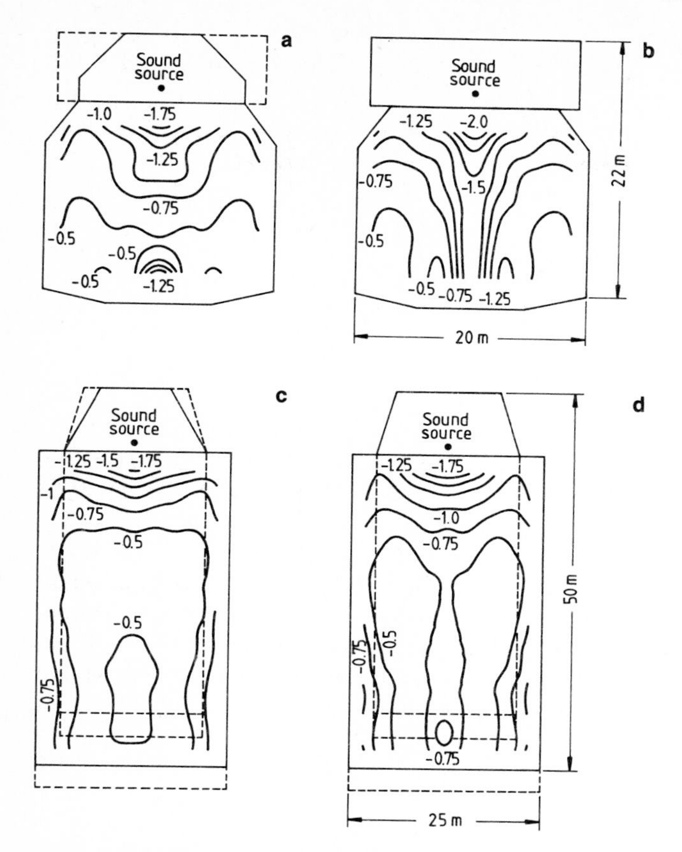

Fig. 6.10 a – d. Contour lines of total subjective preference with the four physical parameters for listeners (Music motif B). **(a)** The Auditorium at Kobe University with well designed side reflectors on the stage; **(b)** the auditorium without side reflectors on the stage; **(c)** the Boston Symphony Hall with side reflector on the stage optimized; **(d)** the Symphony Hall with the original side reflectors on the stage

In addition, reflections from the side reflectors on the stage may be produced for listeners getting small values of IACC, because of its unimportance for performers [6.12]. If the side reflectors on the stage are arranged as shown in Fig. 6.10a, c, then the total scale values of preference for listeners are much improved, particularly in the seating areas close to the stage and the center seats. Figure 6.10b, d gives results for existing halls, the Auditorium at Kobe University and the Symphony Hall in Boston, respectively. Balconys and proscenium of the Boston Symphony are taken into account in the simulation. Since the original shape of the Boston hall and the stage enclosure are well designed, the degree of improvement by stage side reflectors is less than that in the case of the Kobe Auditorium.

As shown in Fig. 6.10a, c, the proportion of preferred areas in the Boston Symphony Hall with more than -0.5 value are much greater than those in the Kobe Auditorium. This is due to the reflections of the side wall and the balcony system. Therefore, it is no exaggeration to say that side walls near to the sound sources may be designed for the listeners, and the ceiling and the back wall in the stage enclosure may be designed for the performers on the stage.

6.4 Concert Halls

Based on results of subjective preference judgments and the auditory physiological data, an asymmetric model of the auditory-brain system has been proposed. It has been interpreted that the spatial and the temporal factors are not interconnected in subjective preference judgments. In fact, the dynamic presentation of spatial sound stimuli arouse the right hemisphere, and this may be identified as the dominant site of the spatial sensation. On the other hand, evidence appeared for left hemisphere superiority in detecting speech and melody with complex temporal or sequential features, which may be represented by the autocorrelation function.

The remarkable convenience in designing auditoria is that the room shape is first determined by only the spatial criterion (IACC) and then its dimension and the absorption characteristics are taken into consideration according to the design range of the effective duration of the autocorrelation function of source signals to be performed (Fig. 5.5b).

In the 1930s *Békésy* made an interesting observation [6.8]: he and most observers and musicians agreed that the musical quality of a courtyard was much better than in a concert hall where the orchestra usually played. In the courtyard, there were no reflections from above and from a portion of one side (an asymmetrical shape). His experience of increased preference may be explained by the interaural cross correlation acting as a binaural-spatial criterion. Another example of the importance of the binaural criterion is indicated in the well-established reputation of certain concert halls, such as the Grosser Musikvereins-Saal in Vienna and the Symphony Hall in Boston [6.13], which have nearly square cross sections producing strong lateral reflections at most seats [6.14–16].

A design study with an asymmetrical plan was performed by an architectural student at Kobe University as an individual exercise, Fig. 6.11. The ceiling, rear walls as well as the stage wall are made up of the diffusing surfaces. The values indicated in the figure are calculated IACCs using (3.7), for which several early reflections ($n \geqq 4$) were considered. The values needed are listed in Table D.1a. For sound fields in concert halls, the maximum values of the interaural cross correlation are defined by (3.8) and are usually found at $\tau = 0$. Therefore, the IACC can be calculated by using correlation values at $\tau = 0$

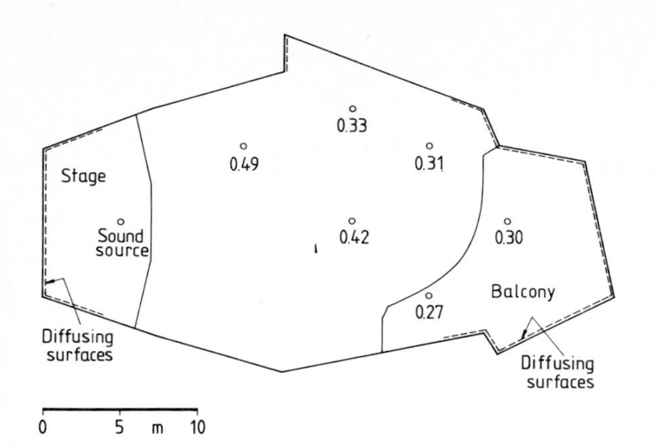

Fig. 6.11. Design study of concert hall by Ko Isogimi, 1979. Values are IACCs calculated using (3.7, 8)

only. For a hall of this shape, the IACCs indicate high preference (IACC smaller than 0.5) at all positions under consideration. Furthermore, three other parameters, i.e., the listening level, the initial time-delay gap, and the reverberation time, are calculated in the exercise, but are not discussed here. In such a way, a variety of improvement in designing concert halls is proposed.

Among them, the Vega-Hall, Takarazuka, for example, has a high ceiling and a balcony on side walls to get enough reflections centered on $\pm 55°$ to listeners [6.17], and the Pikes Peak Center, Colorado Spring, with tilt side walls and movable stage towers has been designed by R. Johnson [6.18]. Calculating the scale value of subjective preference at each seat, the Washington Center, Olympia, was designed in an asymmetrical shape [6.19].

If a music motif to be performed is classified by the effective duration of its autocorrelation function, we can then design the optimum dimensions of the concert hall. But there is a drawback: even for a given music motif, the short-time autocorrelation function changes during the piece of music. However, the effective durations of the autocorrelation functions are often limited to a certain range. This may be related to the fact that during composition, the composer perhaps imagines a certain type of concert hall in which the music is to be played. For example, organ music is composed to be played in churches, and chamber music is intended to be performed in small halls.

For multiple-purpose concert halls, the optimal conditions are strongly concerned with temporal-monaural criteria. To approach optimal conditions one must

1) adjust the delay and amplitude of early reflections by electro-acoustic equipment and control the absorption coefficient of side walls; and
2) control reverberation time by using reverberators [6.20] and by changing acoustic materials of several boundaries in the hall.

7. Acoustic Test Techniques for Concert Halls

After construction, acoustic measurements are usually performed in concert halls to test whether the objective quantities for the sound fields in the concert halls match the quantities designed. The data accumulated here may be used for future designing of concert halls.

7.1 Transfer Function or Impulse Response Measurements

7.1.1 Single Pulse Method

Let us consider a single boundary surface, and a single-pulse sound source $s(t)$ located in front of the surface. The sound pressure $f(t)$ at a receiving point may then be expressed in the form

$$f(t) = A_0 s(t) + A_1 s(t) * w_1(t - \Delta t_1)\,, \qquad \text{where} \tag{7.1}$$

$$s(t) = \begin{cases} 1, & 0 \leqq t \leqq \sigma \\ 0, & \text{otherwise} \end{cases} \tag{7.2}$$

and A_1 is determined by the $(1/r)$ law and $A_0 = 1.0$. The first and second terms of the right-hand side of (7.1) are separated by the time delay Δt_1.

Fourier transforming both terms gives

$$S(\omega) \quad \text{and} \quad S(\omega) A_1 W(\omega)\,. \tag{7.3}$$

The ratio of both terms supplies the transfer function, because A_1 is known.

Figure 7.1 illustrates measurements of the transfer function for the reflection from a single rectangular plate. In this figure the values calculated with (2.19) are also plotted.

The dimensions of the plate are: $a = 30$ cm and $b = 45$ cm. The distances are $l_0 = m_0 = 212$ cm with the angles of incidence and reflection $\alpha_i = \alpha_r = 45°$. During the measurements, a single impulse with a duration of 20 μs was supplied to a loudspeaker. The transfer function can be measured in the range of the energy spectrum of the source. This technique can be applied to measuring the transfer function of each wall in a concert hall. Another example is the measurement of the head-related transfer function, Sect. 2.5.1.

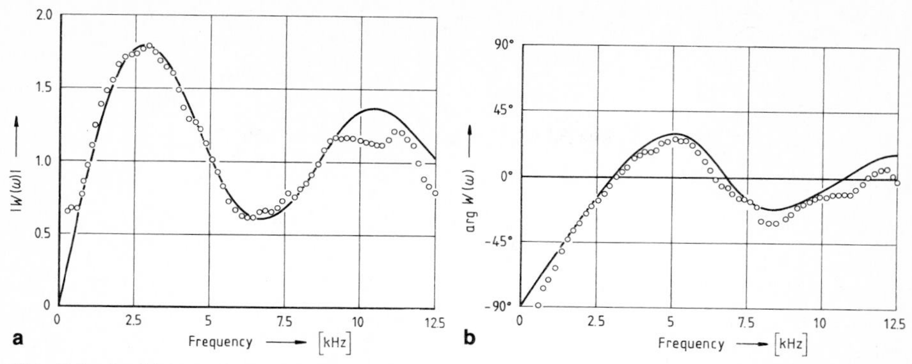

Fig. 7.1a, b. Measured and calculated transfer function $W(\omega)$ for reflection from a single rectangular plate (see Fig. 2.3): $a = 30$ cm, $b = 45$ cm; $r_0 = (150$ cm, 150 cm, 0 cm$)$, $r = (-150$ cm, 150 cm, 0 cm$)$, so that $m_0 = l_0 = 212$ cm. (———) Theoretically calculated values (Sect. 2.2); (○ ○ ○) measured values obtained by the single-pulse method. **(a)** Amplitudes, $|W(\omega)|$; **(b)** phases, arg $W(\omega)$

7.1.2 Correlation Method

The correlation method of measuring the impulse response of an acoustic linear system is illustrated in Fig. 7.2. Let $s_i(t)$ be the input signal to the system and $s_0(t)$ be the output signal. Then the cross correlation between the input and output signals can be expressed by

$$\begin{aligned}\Phi_{io}(\tau) &= \lim_{2T\to\infty} \frac{1}{2T} \int_{-T}^{+T} s_i(t) s_o(t+\tau)\, dt \\ &= \int_{-\infty}^{+\infty} h(\nu)\, \Phi_{ii}(\tau-\nu)\, d\nu \\ &= h(t) * \Phi_{ii}(t)\,,\end{aligned} \tag{7.4}$$

where $s_o(t) = s_i(t) * h(t)$, $h(t)$ being the impulse response of the linear system to be determined and $\Phi_{ii}(\tau)$ is the autocorrelation function of the input signal.

If a white noise signal with the constant power spectrum N is supplied to the system, then

$$\Phi_{ii}(\tau) = N\delta(\tau)\,. \tag{7.5}$$

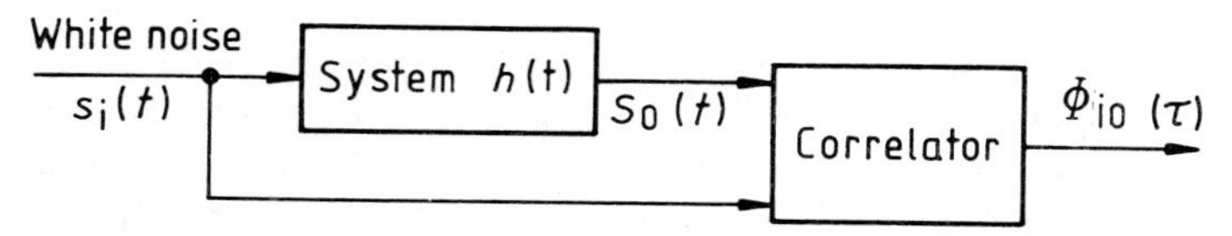

Fig. 7.2. Correlation method of measuring the impulse response

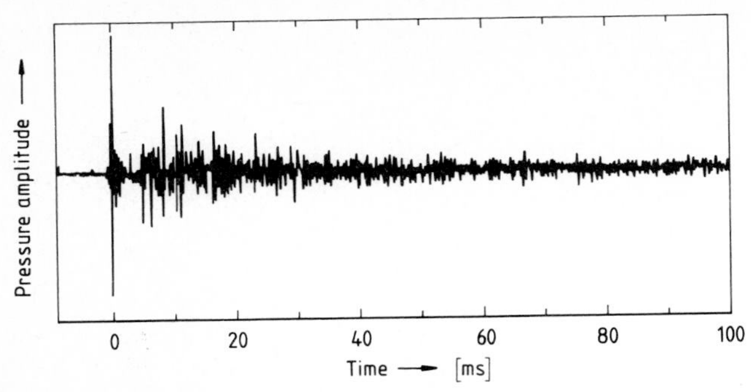

Fig. 7.3. Impulse response in a room with white noise measured by correlation method. Bandwidth of the test signal and the response: 10 kHz

Thus, (7.4) becomes

$$\Phi_{io}(\tau) = Nh(\tau) . \tag{7.6}$$

Therefore, we can obtain the impulse response of the system $h(\tau)$.

If there is another kind of sound source acting as a disturbance $d(t)$ in the hall simultaneously, then the output signal becomes

$$s_0(t) = s_i(t) * h(t) + d(t) * g(t) , \tag{7.7}$$

where $g(t)$ is the impulse response between the disturbing source and the observation.

The cross correlation function is thus expressed by

$$\Phi_{io}(\tau) = \int_{-\infty}^{+\infty} h(v)\, \Phi_{ii}(\tau - v)\, dv + \int_{-\infty}^{+\infty} g(v)\, \Phi_{id}(\tau - v)\, dv , \tag{7.8}$$

where $\Phi_{id}(\tau)$ is the cross correlation between the signals $s_i(t)$ and $d(t)$.

Since $s_i(t)$ and $d(t)$ are independent, the second term becomes zero, under the condition that one of the average values $\overline{s_i(t)}$ or $\overline{d(t)}$ is zero. Thus (7.8) becomes identical to (7.6).

An example of measuring the impulse response in a classroom is shown in Fig. 7.3. In this system the transfer function of the loudspeaker is also included.

7.1.3 Fast Method Using the Pseudo-Random Binary Signal

Alrutz [7.1] proposed a fast method, using a pseudo-random binary signal, to take acoustic measurements in rooms. First of all, we shall consider the binary signal generated by a system with a shift register as shown in Fig. 7.4 [7.2]. When $n = 4$ and $k = 3$ and also initially shift registers contain -1, then the output sequence m_j becomes:

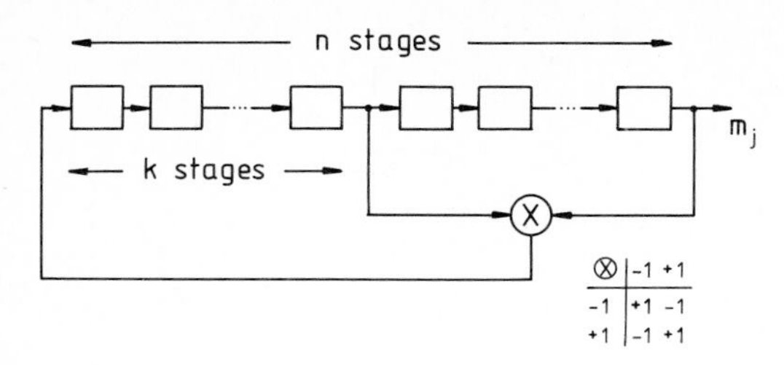

Fig. 7.4. Generation of pseudo-random noise [7.2]

$$-1,\ -1,\ -1,\ -1,\ +1,\ +1,\ +1,\ -1,\ +1,\ +1,\ -1,\ -1,\ +1,\ -1,\ +1\,.$$

This is repeated with a period of 15 binary digits. The largest possible period for the system is given by $L=2^n-1$, and it is called a maximum-length sequence. The maximum-length sequence has the following properties:

$$\sum_{j=0}^{L-1} m_j = -1\,, \qquad \text{and} \tag{7.9}$$

$$\Phi_l = \sum_{j=0}^{L-1} m_j m_{j+l} = \begin{cases} L, & \text{for } l = 0 \bmod L \\ -1, & \text{otherwise}\,. \end{cases} \tag{7.10}$$

The autocorrelation function given by (7.10) shows that the value at the origin is L, and -1, otherwise. From these properties, the maximum-length sequence is called the pseudo-random signal.

Next, consider the single pulse $s(t)$ defined by (7.2). If the duration σ is short enough, then

$$S(\omega)=1\,, \qquad \text{for} \qquad 0 \leqq \omega \leqq \Omega_2\,,$$

where Ω_2 is the upper limit of the audible frequency [rad/s]. Such a signal is applied to a linear system under investigation. Then the output signal corresponds to the impulse response for the frequency range required.

Let us now consider a signal including the maximum-length sequence defined by

$$x(t) = \sum_{j=-\infty}^{\infty} m_j s(t-\mathrm{j}\sigma)\,. \tag{7.11}$$

When the signal is applied to the linear system, then the output signal yields

$$y(t) = \sum_{j=0}^{L-1} m_j h[(t-\mathrm{j}\sigma)_{\mathrm{mod}\,T}]\,, \tag{7.12}$$

where $T=\sigma L$. It is imposed that $h(t)=0$ for $t>T$.
The data sampled from $y(t)$ at every σ interval may be written by

$$y_k = \sum_{j=0}^{L-1} m_j h[\sigma(k-j)_{\mathrm{mod}\,L}]\,, \qquad k = 0, 1, \ldots, L-1$$

$$= \sum_{j=0}^{L-1} h_j m_{k-j}\,, \tag{7.13}$$

where $h_j = h(\sigma j_{\mathrm{mod}\,T})$.
If we set

$$y = \begin{bmatrix} y_0 \\ y_1 \\ \vdots \\ y_{L-1} \end{bmatrix} \quad \text{and} \quad h = \begin{bmatrix} h_0 \\ h_1 \\ \vdots \\ h_{L-1} \end{bmatrix}, \tag{7.14}$$

then (7.13) becomes

$$y = Mh\,, \qquad \text{where } M = [m_{ij}] \text{ and} \tag{7.15}$$

$$m_{ij} = m_{i+j-2\,\mathrm{mod}\,L}\,. \tag{7.16}$$

To obtain the impulse response h, we find the inverse matrix of M

$$\tilde{M} = [\tilde{m}_{ij}]\,, \qquad \tilde{m}_{ij} = m_{ij} - 1 \tag{7.17}$$

such that

$$\tilde{M}M = (L+1)\,I_L\,,$$

where I_L is the $L \times L$ unit matrix. Thus, the impulse response is derived by

$$h = \frac{1}{L-1}\tilde{M}y\,. \tag{7.18}$$

L+

To enable fast computation, we introduce a $(L+1) \times (L+1)$ matrix defined by

$$U = \begin{bmatrix} 1 \ldots 1 \\ \vdots\; M \\ 1 \end{bmatrix} = \begin{bmatrix} 0 \ldots 0 \\ \vdots\; \tilde{M} \\ 0 \end{bmatrix} + \begin{bmatrix} 1 \ldots 1 \\ \vdots \quad \vdots \\ 1 \ldots 1 \end{bmatrix} \tag{7.19}$$

so that (7.18) is rewritten as

$$\begin{bmatrix} 0 \\ h \end{bmatrix} = \frac{1}{L+1}\left[U\begin{pmatrix} 0 \\ y \end{pmatrix} - \begin{bmatrix} \bar{y} \\ \vdots \\ \bar{y} \end{bmatrix} \right], \qquad \text{where} \tag{7.20}$$

$$\bar{y} = \sum_{j=0}^{L-1} y_j\,.$$

According to *Lempel* [7.3], a permutation matrix P exists to express

$$U = P^T H_{2^n} P \, ,$$

where H is a Silvester-type Hadamard matrix, each element being either 0 or 1. It has the symmetrical property as given by

$$H_{2^n} = \begin{bmatrix} H_{2^{n-1}} & -H_{2^{n-1}} \\ -H_{2^{n-1}} & -H_{2^{n-1}} \end{bmatrix} \quad (7.21)$$

$$H_1 = (1) \, .$$

This algorithm enables us to compute the impulse response only by a summation of the output data y from the linear system, without any multiplication operation, as shown in Fig. 7.5. A computational program is described in Appendix E. The method is particularly useful in obtaining the ears' impulse responses $h_l(t)$ and $h_r(t)$ for each seat in a concert hall. After getting the

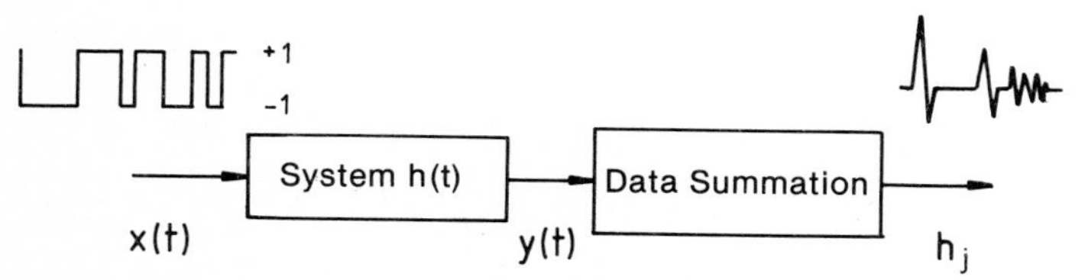

Fig. 7.5. Principle of the fast method of measuring the impulse response

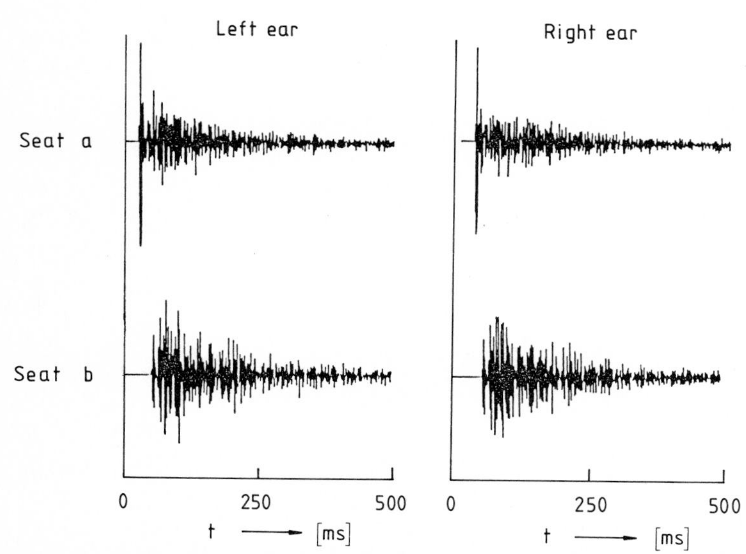

Fig. 7.6. Impulse responses in the Auditorium, Kobe University at both ears obtained with the "maximum-length" pseudo-random noise as test signal, $L = 8191$ and $\sigma = 333$ μs [7.4]

two impulse responses, the optimum objective parameters can be easily analyzed. Typical examples of the impulse responses at two seats in the Auditorium at Kobe University are shown in Fig. 7.6, in which the location of the two seats are indicated by a and b in Fig. 7.9 [7.4].

7.2 Reverberation

The reverberation-measurement method described here was developed by *Schroeder* [7.5, 6]. In the traditional method of measuring the reverberation time, the stationary white noise $n(t)$ is radiated into a concert hall, after passing through a filter (octave or third octave band). Having reached a "steady state", the filtered noise is switched off at $\tau=0$ to observe the reverberation process in the hall, so that $n(\tau)=0$ for $\tau \geqq 0$. The signal received at a receiving point r is expressed by

$$s(t) = \int_{-T}^{0} n(\tau)h(t-\tau)d\tau\,, \tag{7.22}$$

where $h(t)$ is the impulse response of the system consisting of the noise filter, transducers and enclosure between the source point r_0 and the receiving point r. To get the steady-state condition before switching off the noise, the filtered noise should be applied for a sufficiently long time $T \to \infty$.

The square of the signal $s(t)$ may be written as

$$s^2(t) = \int_{-T}^{0} d\tau \int_{-T}^{0} d\theta\, n(\tau)n(\theta)h(t-\tau)h(t-\theta)\,. \tag{7.23}$$

Averaging (7.23) over an ensemble of noise signals gives

$$\langle s^2(t)\rangle = \int_{-T}^{0} d\tau \int_{-T}^{0} d\theta N\delta(\theta-\tau)h(t-\tau)h(t-\theta) = N\int_{-T}^{0} h^2(t-\tau)d\tau\,, \tag{7.24}$$

where $\langle n(\tau)n(\theta)\rangle = N\delta(\theta-\tau)$.
Setting $x = t-\tau$,

$$\langle s^2(t)\rangle = N\int_{t+T}^{t} h^2(x)dx\,. \tag{7.25}$$

Therefore, the ensemble average of the squared noise decay $\langle s^2(t)\rangle$ is obtained by a certain integral over the squared impulse response $h^2(x)$. An example of decay curve measurements is shown in Fig. 7.7 [7.7].

After getting impulse response either $h^2(t)$ or $h(t)$, the energy ratio between the direct sound early-plus-subsequent reverberation, for example is obtained by

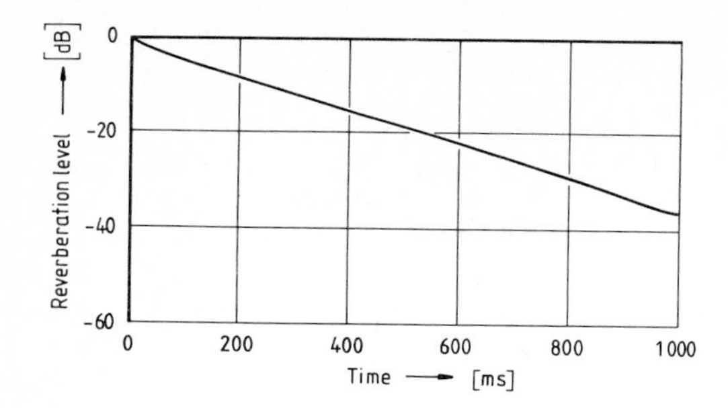

Fig. 7.7. Integrated decay curve obtained from the impulse response by squaring and integrating. Effect of background noise becomes visible only above 950 ms [7.7]

$$A^2 = \frac{\int_{\varepsilon}^{\infty} h^2(\tau)\,d\tau}{\int_{0}^{\varepsilon} h^2(\tau)\,d\tau}, \tag{7.26}$$

where ε is the small delay time for the duration of the direct sound. Furthermore, any other ratio may be obtained, such as

$$D(\Delta t_n) = -\frac{\int_{0}^{\Delta t_n} h^2(\tau)\,d\tau}{\int_{0}^{\infty} h^2(\tau)\,d\tau}. \tag{7.27}$$

When $\Delta t_n = 50$ ms, then it is called the "Deutlichkeit" as defined by *Thiele* [7.8]. And, if $\Delta t_n = \varepsilon$, then $D = 1/(1 + A^2)$.

7.3 Interaural Cross Correlation

The interaural cross correlation can be measured, either with a dummy head or with a human head, by setting small microphones at the entrances of the ear canals. The total transfer function of the microphones and the amplifiers should be adjusted to the ear sensitivity. Then the output signals are fed into a correlator or digitized and Φ_{lr} computed.

The source signals listed in Table 2.1 and a filtered Gaussian noise may be used to compare the calculated with the measured values. An example of this comparison for a certain seat in an existing concert hall is shown in Fig. 7.8. In the figure, the calculated values of the IACC are plotted according to the number of early reflections. The value $n = 0$ indicates the IACC of the direct sound only. If we add the reflections arriving from side walls, then the IACC rapidly decreases, as can seen from the calculated values at $n = 1$ and $n = 2$. In this case, the calculated values of the IACC almost converge, when $n \geqq 4$.

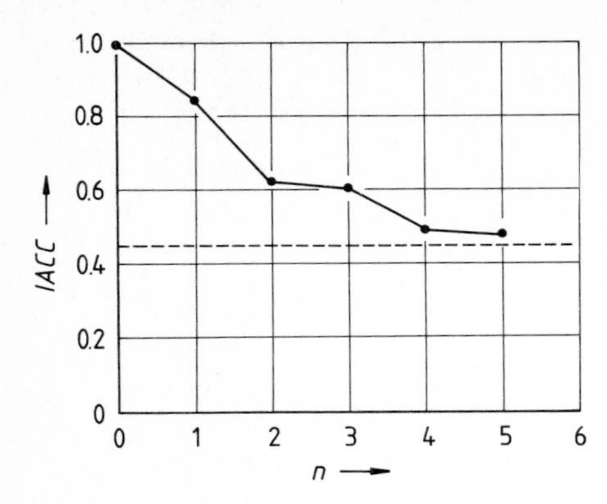

Fig. 7.8. Convergence of the calculated values (●) of IACC according to the number of early reflections at a seat in a concert hall with the measured value (– – –) using a dummy head. Music motif A was reproduced at the center of stage by a loudspeaker with sufficient nondirectivity below 2 kHz

The measured value of the IACC at the seat included all reflections $n \to \infty$, so that the values are slightly smaller than those calculated, as indicated by the dashed line in Fig. 7.8. The correlation coefficients between measured and calculated values of the IACC for concert halls were greater than +0.86, when music motifs A and B were used.

From the impulse responses obtained from both ears, we may calculate the interaural cross correlation similarly to (5.1). After analyzing the impulse response for both ears $h_{l,r}(t)$, for example, by the fast method (Sect. 7.1.3), the IACC may be simply obtained by [7.9]

$$\Phi_{h_l h_r}(\mathrm{j}\sigma) = \sum_{k=0}^{L-1} h_{l,k} h_{r,j+k}\,, \qquad \Phi_{h_l h_r}(-\mathrm{j}\sigma) = \sum_{k=0}^{L-1} h_{r,k} h_{l,j+k}\,,$$

$$\Phi_{h_l h_l}(\mathrm{j}\sigma) = \sum_{k=0}^{L-1} h_{l,k} h_{l,j+k}\,, \tag{7.28}$$

$$\Phi_{h_r h_r}(\mathrm{j}\sigma) = \sum_{k=0}^{L-1} h_{r,k} h_{r,j+k} \qquad \text{and}$$

$$\Phi_{lr} = \Phi_{h_r h_l} * \Phi_p\,, \qquad \Phi_{ll} = \Phi_{h_l h_l} * \Phi_p\,, \qquad \Phi_{rr} = \Phi_{h_r h_r} * \Phi_p\,, \tag{7.29}$$

$$\phi_{lr}(\pm\mathrm{j}\sigma) = \frac{\Phi_{lr}(\pm\mathrm{j}\sigma)}{\sqrt{\Phi_{ll}(0)\,\Phi_{rr}(0)}}\,, \tag{7.30}$$

Φ_p being the autocorrelation function of the source signal including the effect of ear sensitivity (Sect. 5.1). For practical convenience, Φ_p is selected as a typical autocorrelation function of a certain bandpass noise, so that (7.30) may be simply calculated. For the frequency range 100 – 500 Hz, contour lines of equal IACC values measured in the Auditorium at Kobe are shown in Fig. 7.9. If these IACC contours are compaired with calculated total preferences, as shown in Fig. 6.10b, we can imagine how the high IACC values in the center line of audience area affect the preference.

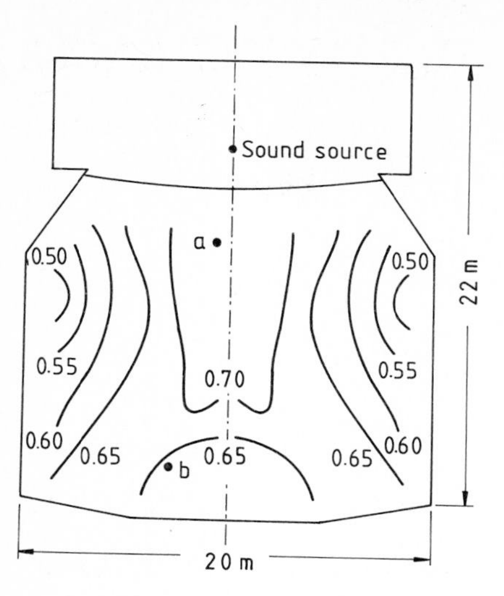

Fig. 7.9. Contour lines of equal IACC value measured in the Auditorium at Kobe University with "maximum-length" signal [7.4]. *a, b*: Seat locations for impulse responses shown in Fig. 7.6

7.4 Subjective Preference Judgments of Sound Fields in Existing Concert Halls

In European concert halls, music motif E (Mozart) was played on the stage and recorded through two microphones installed at both ears of a dummy head with realistic surface impedance for the ear canal and pinnae [7.10]. Using modified reproduction filters from that described in Sect. 3.2 [7.11], the music sound fields were reproduced for paired comparison tests. The listening level was adjusted to peak values of 80 dBA which almost corresponds to the preferred level.

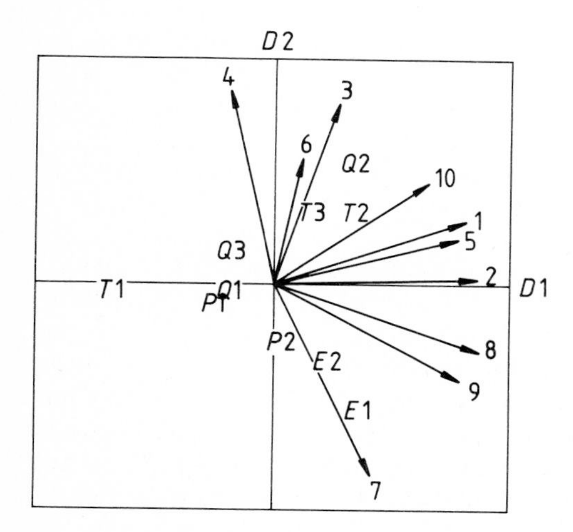

Fig. 7.10. Preference space for 10 seats in 4 concert halls with reverberation time less than 2.2 s. Dimension D1 is called "consensus preference", and D2 reflects individual preference differences (10 sound fields and 10 German subjects) [7.10]. *E, P, Q, T* denote sound fields from corresponding concert halls; numbers 1 – 10 denote subjects

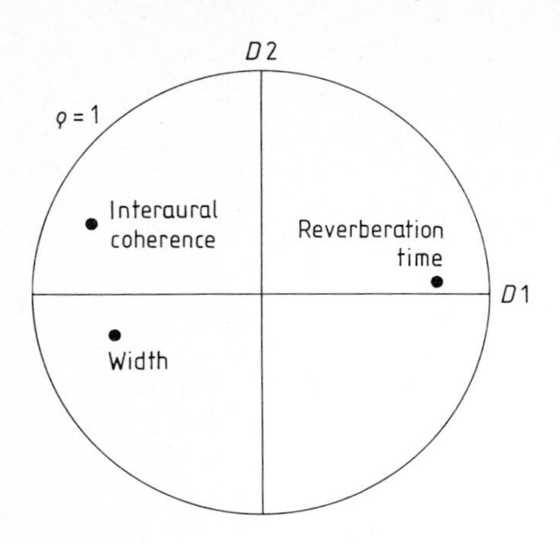

Fig. 7.11. Correlation of three objective parameters with dimensions D1 and D2 shown in Fig. 7.10. The IACC has the highest negative correlation with D_1, "consensus preference"

The individual preference data were analyzed by a linear factor analysis [7.12]; and, the results are shown in Fig. 7.10. The sound fields recorded in each of several concert halls are signified by capital letters followed by numbers. Vectors representing 10 different listeners in the figure indicate that they have different musical preferences. Nevertheless, all vectors were directed to the right in the two significant preference dimensions except for that of listener 4. In this sense, the dimension D_1 is called "consensus preference". For example, we can see that sound field E_1 is the most preferred sound field (or seat) for listener 7 and sound field Q_2 is preferred for listener 3. The sound field T_1 was the least preferred for all listeners. Thus, the dimension D_2 reflects the individual differences of the listeners.

Similar significant parameters obtained for sound fields may be found in the next step of the consensus preference correlations. Among the geometric and acoustic parameters examined by factor analysis, and which underly these preference judgments, are the interaural coherence, the reverberation time and the width of the halls, Fig. 7.11. In this case, the most significant parameter found is interaural coherence. As is clear from the figure, higher interaural coherence causes smaller consensus preference. This agrees well with the results in Sect. 4.3. Also, the subjective preference for each sound field is correlated with reverberation time. If only the reverberation times below 2.2 s are considered, then the longer reverberation time is preferred.

The preferred reverberation time obtained in Sect. 4.3 for the Mozart motif is 1.2 s, which is much shorter than the value 2.2 s. There are two possible reasons for the discrepancy:

1) only the fast passage of the music (5 s) was used in Sect. 4.3 while a longer passage (about 20 s) was used in this investigation;
2) energy of the subsequent reverberation or the total amplitude of reflections (A) might be smaller in real concert halls, so that the difference of the scale values of preference of 1.2 s and 2.2 s sound fields could be small as well.

Another physical measure showing strong negative correlation with consensus preference appears to be the width of the halls. Although the correlation coefficient between the values of interaural coherence and the reverberation time was -0.15, the correlation coefficient between the width and the interaural coherence was significantly high, i.e., 0.86, and the coefficient between the width and the reverberation time was -0.62. Therefore, the width itself is not considered as an independent objective parameter, and the preference in this investigation may be explained by only the interaural coherence and the reverberation time. It is worth noting that the initial time-delay gap was limited in a certain range, because recording positions were selected only around the center of the halls, and the listening level was kept at a constant level, so that these two parameters could not appear as significant factors in the subjective judgments.

For wide halls the first strong sound may arrive at the listener's position from the ceiling. These ceiling-reflected sounds produce similar signals at both ears. Thus, wider halls generally have higher magnitude of IACC, meaning lower preference (Fig. 7.9).

Of particular interest is another approach using a diagnostic system to obtain the subjective preference at each seat in existing concert halls. After getting the four objective parameters as described in Sect. 7.1, the scale values of preference may be calculated by (5.11 – 19). This is possible without any subjective judgments by listeners, if the autocorrelation function of the source signal and the most preferred listening seat concerning its level are known.

A diagnostic system of sound quality in existing rooms is shown in Fig. 7.12. In a concert hall under test, the M-sequence sound signal is produced

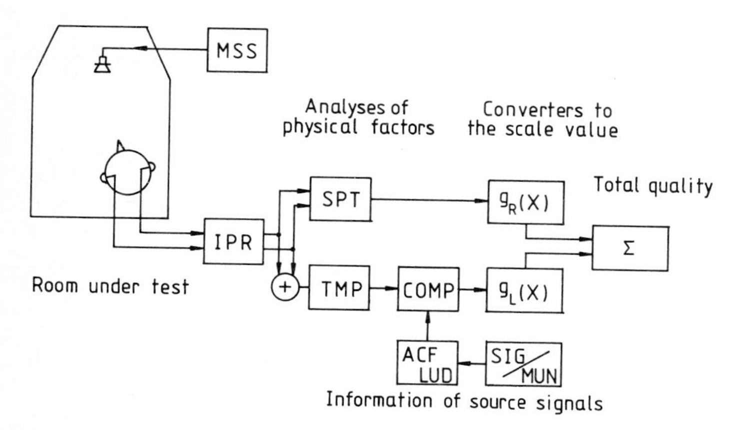

Fig. 7.12. A diagnostic system analyzing the four physical factors and evaluating the sound quality at each seat in a existing room. MSS: M-Sequence signal; IPR: Impulse response analyzers; SPT: Spatial factor (IACC); TMP: Temporal factors (LL, Δt_1, T_{sub}; A); $g_R(X)$: Spatial factor to scale-value converter; $g_L(X)$: Temporal factors to scale-value converters; Σ: Total quality calculator; COMP: Comparators; ACF, LUD: Autocorrelation function or τ_e, Loudness information from musical note; SIG/MUN: Source signals and/or musical notes

from a loudspeaker located on the stage. The acoustic signals at the ears of a dummy head or a real head are picked up by two microphones. Output signals of the microphones are fed into a personal computer through an analog-to-digital converter. Then the impulse responses given by Eq. (7.20) are calculated by applying the *M*-sequence to the Hadamard matrix conversion algorithm with only addition and subtraction operations. From the two impulse responses, the spatial factor and the temporal factors are easily analyzed.

After getting the physical factors, the total scale value of preference may be obtained by the transformation formulas and a linear summation due to the theory of preference, which consists of the logarithmic-monaural detectors, the linear-binaural detector, the preference evaluator based upon the 3/2 power process, and the independent action of each physical factor on the subjective preference space. In analyzing the scale value, the optimal conditions of the temporal factors which can be determined by the sound source information (loudness information, for example, in music notation: *mf*, *p*, <; and the envelope of the autocorrelation function) are compared with physical factors analyzed (level of listening, Δt_1 and T_{sub}). The autocorrelation function of each music motif could be approximately analyzed by the use of a music synthesizer. More conveniently, a possibility exists that the envelope of the autocorrelation function or τ_e could be directly calculated from the musical score considering the envelope and intensity of each note for a given instrument.

Appendices

A. Subjective Diffuseness

The IACC magnitude may be related to subjective diffuseness or spatial impression as well as to the subjective preference.

As shown by arrows in Fig. A.1, *Damaske* [A.1] arranged four loudspeakers in an anechoic chamber, and wide band noises with different coherence magnitudes were fed into the loudspeakers (2.6 m from the center of a listener). Results of subjective diffuseness are indicated as black shadows in the figure. Obviously, subjective diffuseness increases with decreasing magnitude of coherence $\bar{k}$ between signals supplied to loudspeakers which may be related to the magnitude of the IACC. Similar results were also obtained by *Keet* [A.2].

Next, let us discuss a more direct relationship between the IACC magnitude and the degree of spatial impression. *Barron* and *Marshall* [A.3] con-

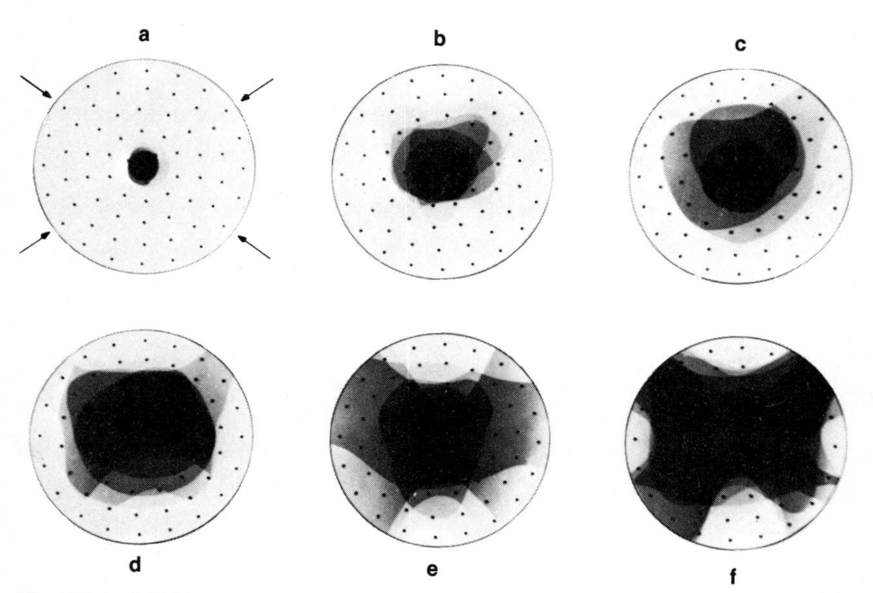

Fig. A.1a – f. Subjective diffuseness with coherence $\bar{k}$ between signals supplied to loudspeakers located at arrows. The signals are wideband white noise 250 Hz – 2 kHz [A.1]. **(a)** $\bar{k} = 0.98$; **(b)** $\bar{k} = 0.72$; **(c)** $\bar{k} = 0.46$; **(d)** $\bar{k} = 0.35$; **(e)** $\bar{k} = 0.25$; **(f)** $\bar{k} = 0.15$

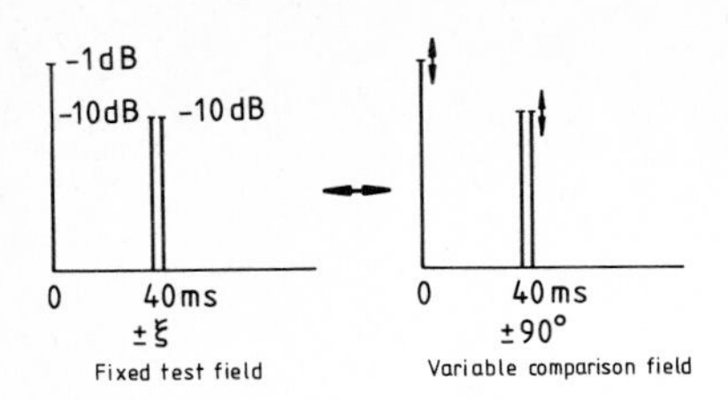

Fig. A.2. Impulse responses of sound fields to compare degree of spatial impression [A.3]

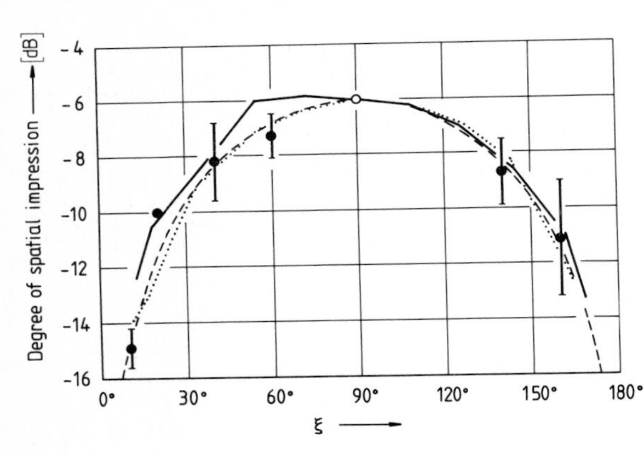

Fig. A.3. Degree of spatial impression for pairs of reflections as the reflection horizontal angle ξ is varied. ⧫: Mean and 90% confidence limit of the mean of experimental results [A.3]; (———) calculated results using (A.3) with measured correlation values indicated in Table D.1e; (·········) calculated results using (A.3) with correlation values calculated using (A.4). (– – –) predicted results [A.3]

ducted a self-testing comparison experiment with Mozart music (music motif E) for sound fields with impulse responses as shown in Fig. A.2. The subjects were asked to adjust the variable sound field for equal spatial impression to the fixed field and the experimenters measured the reflection amplitude with reference to the direct sound amplitude in dB. The measured results of the reflection amplitude as the degree of spatial impression are plotted in Fig. A.3 as a function of the reflection horizontal angle. The value at $\xi = 90°$ is -6 dB by definition.

For both sound fields, the IACC magnitude may be calculated by (3.7). The IACC of the fixed sound field is obtained at $\tau = 0$, so that [A.4]

$$\phi_{lr}^{(\pm\xi)}(0) = \frac{1 + \frac{1}{4}\Phi_{lr}^{(\xi)}(0)}{1 + \frac{1}{8}[\Phi_{ll}^{(+\xi)}(0) + \Phi_{ll}^{(-\xi)}(0)]}\,, \tag{A.1}$$

where we set $A_1^2 = A_2^2 = \frac{1}{8}\,(-9\text{ dB})$,

$$\Phi_{lr}^{(0)}(0) = \Phi_{ll}^{(0)}(0) = \Phi_{rr}^{(0)}(0) = 1\,,$$

$$\Phi_{lr}^{(+\xi)}(0) = \Phi_{lr}^{(-\xi)}(0) = \Phi_{lr}^{(\xi)}(0)\,, \quad \text{and} \quad \Phi_{ll}^{(\pm\xi)}(0) = \Phi_{rr}^{(\mp\xi)}(0)\,.$$

On the other hand, the IACC of the variable sound field is

$$\phi_{lr}^{(\pm 90°)}(0) = \frac{1 + 2A_1^2\Phi_{lr}^{(90°)}(0)}{1 + A_1^2[\Phi_{ll}^{(+90°)}(0) + \Phi_{ll}^{(-90°)}(0)]}\,, \tag{A.2}$$

where the amplitude A_1^2 is unknown.

Since the spatial impression of both sound fields is the same, it is assumed that the right-hand side of (A.2) must be identical with $\phi_{lr}^{(\pm\xi)}(0)$ given by (A.1). Thus, the amplitude can be derived from

$$A_1^2 = \frac{1 - \phi_{lr}^{(\pm\xi)}(0)}{\phi_{lr}^{(\pm\xi)}(0)\,[\Phi_{ll}^{(+90°)}(0) + \Phi_{ll}^{(-90°)}(0)] - 2\,\Phi_{lr}^{(90°)}(0)}\,. \tag{A.3}$$

The reflection amplitude defined as the degree of spatial impression may be obtained by $10 \log 2A_1^2$ in dB. The correlation values needed for the calculation using the Mozart motif are indicated in Table D.1 e.

On the other hand, for any sound source with the pressure spectrum S_m ($m = 1, 2, \ldots, M$) analyzed by corresponding bandpass filters as being performed at the peripheral autditory system, the correlation values may be approximately obtained by

$$\begin{aligned} \Phi_{lr}^{(n)}(\tau) &= \sum_{m=1}^{M} S_m^2\,\Phi_{lr}^{(m,n)}(\tau)\,, \\ \Phi_{ll}^{(n)}(0) &= \sum_{m=1}^{M} S_m^2\,\Phi_{ll}^{(m,n)}(0)\,, \\ \Phi_{rr}^{(n)}(0) &= \sum_{m=1}^{M} S_m^2\,\Phi_{rr}^{(m,n)}(0)\,, \end{aligned} \tag{A.4}$$

where $\Phi_{lr}^{(m,n)}(\tau)$, $\Phi_{ll}^{(m,n)}(0)$ and $\Phi_{rr}^{(m,n)}(0)$ are measured correlation values for each bandpass noise (m) and each sound direction (n). This may be operative in the auditory pathways. Correlation values measured with the 1/3 octave bandpass noise are indicated in Table D.2. As space is limited, values of $\Phi_{lr}^{(m,n)}(\tau)$ are listed only at $\tau = 0$ and at an interval of octave. For other frequency bands, if necessary, values are interpolated. Inspecting the values, it may found that the effective angle of reflections required to minimize the IACC for sound fields with frontal direct sound differs according to the frequency range. For a frequency range below 500 Hz, $\xi = 90°$; for 1000 Hz as obtained for music sources, $\xi = 54°$; $\xi = 36°$ for 2000 Hz; and $\xi = 18°$ for 4000 Hz.

Calculated results are also plotted in Fig. A.3. The experimental values obtained by the equal spatial impression agree with the calculated values. In the calculation with (A.4), the measured spectrum indicated in Table A.1 and values in Table D.3 are used. For more general angles (ξ, η), results are shown in Fig. A.4. Agreements between experimental and calculated values are satisfactory. In this calculation, the angle given by $\sin^{-1}(\sin\xi\cos\eta)$ is substituted for the horizontal angle ξ. These results further support the interaural cross correlation model in the auditory pathways described in Sect. 5.1. For a sound with predominantly low frequency (range below 250 Hz), the interaural cross

Table A.1. Measured spectrum of the Mozart motif ($2T = 35$ s), after passing through the A-weighting filter

1/3 oct. band center freq. [Hz]	A_m^2
125	0.011
250	0.166
500	0.984
1000	1.000
2000	0.471
4000	0.089

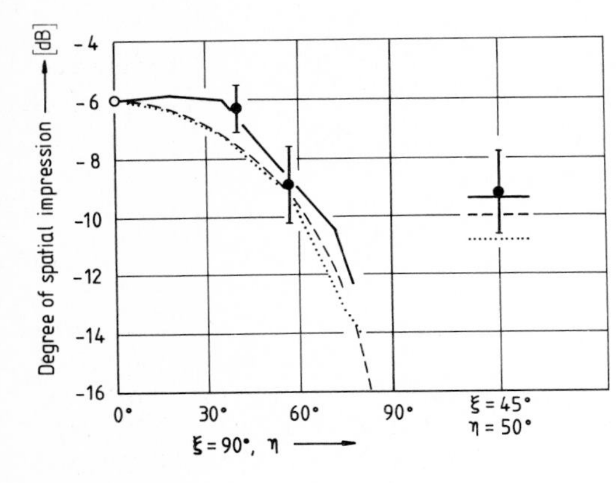

Fig. A.4. Degree of spatial impression for pairs of reflections as the horizontal and elevation reflection angles are varied. ⫯: Mean and 90% confidence limit of the mean of experimental results [A.3]; (———) calculated results using (A.3) with measured correlation values indicated in Table D.1e; (·········) calculated results using (A.3) with calculated correlation values from (A.4); (– – –) predicted results [A.3]

correlation function has no sharp peaks for the delay range of $|\tau| \leqq 1$ ms. Under this condition, we perceive no sharp directional sound, i.e., the diffusing sound.

B. An Example of Individual Difference in Preference Judgment

Individual preference results obtained for sound fields with a single reflection are shown in Fig. B.1 (music motif B) as a function of its delay time. (The integrated data are described in Sect. 4.2.1.) The individual data of the probability of positive judgments is represented by a number in this figure. For example, the most preferred delay time of subjects 5, 8, 9 and ② was at 32 ms and that of subjects 2, 6 and ⓪ was at 64 ms. Only subject 7 preferred the delay time of 6 ms, but this kind of response was infrequent throughout the investigations. All subjects had lowest preference at 256 ms.

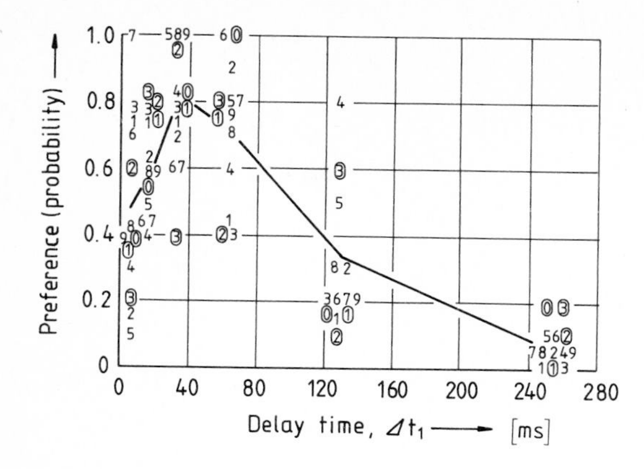

Fig. B.1. Individual preference results as a function of the delay time of a single reflection (13 subjects, denoted by numbers). Subject numbers contained in circles should be added by ten, ex.: ② = 12

Summation of individual probability results shows that the most preferred delay time is at 32 ms, which corresponds to Curve B in Fig. 4.2. Further discussion concerning individual differences is given in Sect. 7.4.

C. Perception of Coloration

The purpose of this supplement is to show further evidence of subjective aspects associated with autocorrelation analysis in the auditory passage.

The perception of coloration was investigated using a sound field composed of a strong primary sound and a delayed weak sound [C.1]. A continuous bandpass noise was used as a source signal in which the autocorrelation function is independent of its time interval and is theoretically analyzable (Fig. C.1). The loudspeaker arrangement which presents the primary sound and the delayed sound is shown at the right in Fig. C.2. The total sound pressure of the two sounds was kept constant at a listener's position in an anechoic chamber (60 dBA). The subject judged the sensation level of the weak delayed sound in comparison to the situation in which only the primary sound was presented. The threshold levels of the weak sound relative to the primary sound as a function of the delay time Δt_1 are shown in the figure. The center frequency of the bandpass filter is 1 kHz. The different symbols in Fig. C.2 signify responses by two subjects.

In a similar manner to (5.4, 5), the threshold level A at a delay time $\Delta t_1 = \tau$ is assumed to be

$$A^{c'} = \frac{1}{k'} |\phi_p(\tau)|_{\text{envelope}}, \tag{C.1}$$

where c' and k' are constants. These constants may be obtained from the results shown in Fig. C.2 as follows:

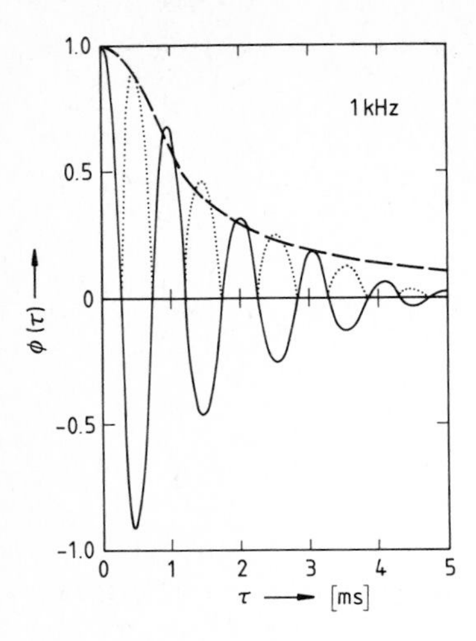

Fig. C.1. Autocorrelation function of continuous bandpass noise with center frequency of 1 kHz. (———) Measured autocorrelation function; (·········) absolute values of the autocorrelation function; (– – –) envelope curve calculated for the noise [C.1]

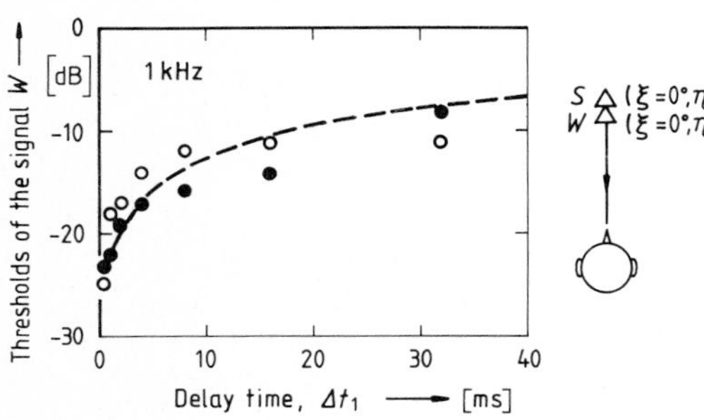

Fig. C.2. Threshold levels of the delayed sound W when the sound S exists as a function of delay time Δt_1 [C.1]. The threshold level at $\Delta t_1 = 0$ is found at -25 dB. Different symbols indicate responses for two subjects; (– – –) denotes calculated values using (C.2)

$$
\begin{aligned}
&c' = -2.0\,, \quad \text{and} \\
&k' = 3.16 \times 10^{-3} \quad \text{or} \quad 10 \log k' = -25\ [\text{dB}]\,,
\end{aligned}
\tag{C.2}
$$

because $20 \log A = -25$ [dB] at $\Delta t_1 = 0$. The dashed curve is the calculated value obtained with (C.1), with the envelope of the autocorrelation function and the constants given by (C.2).

This indicates that the perception of coloration in a certain delay range of a weak sound can be well described in relation to the autocorrelation function of the source signal. This holds also for sound fields with delayed weak sound arriving from the lateral direction. Similar results with 250 Hz and 4 kHz were obtained [C-1].

D. Correlation Functions at Both Ears

To calculate using (3.7, 8) the interaural cross correlation when designing concert halls, theaters and any auditoriums, values of correlations for each sound source are listed in Table D.1.

Table D.1. Correlation functions measured at both ears for each single sound signal arriving as a function of horizontal angle of incidence $\xi(\eta = 0°)$. For the range $180° < \xi < 360°$, the values may be obtained by setting $\xi = 360° - \xi$ and interchanging the suffixes l and r

	Delay τ [ms]	Horizontal angle ξ										
		0°	18°	36°	54°	72°	90°	108°	126°	144°	162°	180°
a) Music motif A (Gibbons)												
	−1.00	−0.43	−0.32	+0.09	+0.17	+0.02	−0.09	0.00	+0.02	0.00	−0.24	−0.37
	−0.95	−0.30	−0.41	−0.02	+0.19	+0.13	+0.11	+0.22	+0.04	−0.02	−0.24	−0.22
	−0.90	−0.06	−0.39	−0.13	+0.17	+0.22	+0.32	+0.37	+0.09	−0.04	−0.19	0.00
	−0.85	+0.09	−0.32	−0.22	+0.15	+0.30	+0.45	+0.45	+0.09	−0.06	−0.11	+0.19
	−0.80	+0.30	−0.19	−0.24	+0.11	+0.30	+0.54	+0.49	+0.06	−0.06	−0.02	+0.30
	−0.75	+0.41	+0.02	−0.22	+0.04	+0.24	+0.54	+0.45	+0.06	−0.04	+0.09	+0.45
	−0.70	+0.43	+0.17	−0.13	0.00	+0.15	+0.45	+0.34	+0.04	−0.02	+0.17	+0.47
	−0.65	+0.34	+0.32	−0.04	−0.04	+0.02	+0.30	+0.22	0.00	+0.02	+0.22	+0.39
	−0.60	+0.19	+0.41	+0.09	−0.09	−0.11	+0.11	0.00	−0.04	+0.04	+0.22	+0.24
	−0.55	−0.06	+0.37	+0.17	−0.09	−0.24	−0.11	−0.15	−0.06	+0.06	+0.15	+0.04
	−0.50	−0.32	+0.28	+0.22	−0.09	−0.32	−0.28	−0.28	−0.06	+0.04	+0.06	−0.17
	−0.45	−0.54	+0.11	+0.22	−0.06	−0.34	−0.39	−0.37	−0.06	+0.02	−0.09	−0.37
	−0.40	−0.65	−0.11	+0.17	−0.02	−0.32	−0.43	−0.39	−0.06	−0.02	−0.22	−0.45
	−0.35	−0.65	−0.32	+0.06	0.00	−0.24	−0.37	−0.30	−0.04	−0.06	−0.32	−0.47
	−0.30	−0.54	−0.47	−0.09	+0.02	−0.09	−0.26	−0.22	0.00	−0.13	−0.41	−0.39
	−0.25	−0.32	−0.58	−0.24	0.00	+0.02	−0.11	−0.06	+0.02	−0.19	−0.41	−0.22
	−0.20	0.00	−0.56	−0.37	−0.02	+0.15	+0.09	+0.06	+0.04	−0.22	−0.34	+0.02
	−0.15	+0.41	−0.43	−0.47	−0.06	+0.24	+0.22	+0.15	+0.02	−0.22	−0.19	+0.26
	−0.10	+0.69	−0.19	−0.49	−0.13	+0.26	+0.28	+0.19	+0.02	−0.17	0.00	+0.52
	−0.05	+0.95	+0.06	−0.43	−0.22	+0.22	+0.26	+0.15	0.00	−0.11	+0.19	+0.65
$\Phi_{lr}(\tau)$	0.00	+0.99	+0.30	−0.32*	−0.32*	+0.09	+0.13	0.00	−0.07	−0.09	+0.30	+0.69
	+0.05	+0.99	+0.52	−0.19	−0.37	0.00	0.00	−0.09	−0.11	−0.04	+0.45	+0.67
	+0.10	+0.92	+0.73	+0.04	−0.34	−0.11	−0.17	−0.22	−0.11	+0.06	+0.52	+0.54
	+0.15	+0.65	+0.82	+0.26	−0.26	−0.26	−0.32	−0.37	−0.13	+0.17	+0.60	+0.39
	+0.20	+0.32	+0.82	+0.47	−0.13	−0.32	−0.47	−0.43	−0.09	+0.26	+0.54	+0.11
	+0.25	−0.04	+0.65	+0.60	+0.04	−0.34	−0.60	−0.47	−0.04	+0.32	+0.43	−0.15

Table D.1 (continued)

	Delay τ [ms]	Horizontal angle ξ										
		0°	18°	36°	54°	72°	90°	108°	126°	144°	162°	180°
	+0.30	−0.32	+0.45	+0.67	+0.22	−0.28	−0.58	−0.41	+0.02	+0.34	+0.26	−0.32
	+0.35	−0.52	+0.19	+0.65	+0.32	−0.17	−0.47	−0.26	+0.09	+0.32	+0.11	−0.47
	+0.40	−0.62	−0.06	+0.54	+0.45	−0.02	−0.26	−0.04	+0.17	+0.28	−0.09	−0.49
	+0.45	−0.60	−0.24	+0.34	+0.52	+0.17	−0.02	+0.22	+0.22	+0.19	−0.17	−0.43
	+0.50	−0.45	−0.39	+0.19	+0.52	+0.34	+0.28	+0.45	+0.26	+0.13	−0.22	−0.28
	+0.55	−0.24	−0.41	0.00	+0.45	+0.49	+0.56	+0.69	+0.28	+0.06	−0.19	−0.11
	+0.60	+0.04	−0.32	−0.11	+0.39	+0.58	+0.77	+0.84	+0.28	+0.02	−0.13	+0.11
	+0.65	+0.28	−0.17	−0.17	+0.28	+0.60	+0.90	+0.90	+0.26	+0.02	+0.04	+0.30
	+0.70	+0.41	0.00	−0.15	+0.17	+0.54	+0.90	+0.86	+0.22	+0.02	+0.09	+0.41
	+0.75	+0.47	+0.22	−0.06	+0.11	+0.43	+0.82	+0.75	+0.17	+0.04	+0.17	+0.45
	+0.80	+0.43	+0.34	+0.04	+0.04	+0.28	+0.62	+0.54	+0.11	+0.09	+0.24	+0.39
	+0.85	+0.28	+0.43	+0.15	+0.02	+0.11	+0.39	+0.30	+0.04	+0.11	+0.24	+0.28
	+0.90	+0.04	+0.41	+0.24	+0.02	−0.04	+0.13	+0.11	0.00	+0.13	+0.19	+0.11
	+0.95	−0.19	+0.34	+0.30	+0.04	−0.17	−0.11	−0.13	−0.02	+0.13	+0.06	−0.11
	+1.00	−0.34	+0.17	+0.30	+0.09	−0.22	−0.24	−0.24	−0.02	+0.11	−0.04	−0.26
$\Phi_{ll}(\tau)$	0.00	+1.00	+0.71	+0.42	+0.32	+0.34	+0.65	+0.62	+0.19	+0.24	+0.52	+0.69
$\Phi_{rr}(\tau)$	0.00	+1.00	+1.12	+1.31	+1.42	+1.27	+1.51	+1.51	+0.75	+0.84	+0.75	+0.71
b) Music motif B (Arnold)												
	−1.00	0.00	−0.07	+0.08	−0.13	−0.03	0.00	−0.04	−0.07	−0.03	−0.06	0.00
	−0.95	+0.17	−0.17	+0.11	−0.07	−0.04	+0.03	−0.03	−0.06	0.00	−0.10	+0.10
	−0.90	+0.17	−0.17	+0.03	+0.01	−0.04	+0.04	−0.03	−0.01	+0.04	−0.10	+0.11
	−0.85	0.00	−0.01	−0.14	+0.07	−0.01	+0.01	−0.04	0.00	+0.01	−0.04	+0.06
	−0.80	−0.17	+0.11	−0.20	+0.06	0.00	−0.01	−0.04	+0.03	−0.03	+0.03	0.00
	−0.75	−0.17	+0.11	−0.14	−0.03	+0.03	−0.03	−0.03	0.00	−0.06	+0.03	+0.01
	−0.70	0.00	−0.01	+0.04	−0.14	+0.01	−0.01	0.00	−0.03	−0.04	0.00	+0.10
	−0.65	+0.24	−0.11	+0.14	−0.17	−0.01	+0.01	0.00	−0.04	−0.01	−0.06	+0.14

	−0.60	+0.24	−0.10	+0.07	−0.07	−0.06	+0.03	0.00	−0.03	+0.01	−0.06	+0.08
	−0.55	−0.04	+0.06	−0.11	+0.07	−0.10	0.00	−0.03	−0.01	0.00	+0.01	−0.07
	−0.50	−0.38	+0.20	−0.20	+0.11	−0.07	−0.06	−0.04	−0.01	−0.01	+0.10	−0.25
	−0.45	−0.48	+0.17	−0.06	+0.01	−0.03	−0.10	−0.07	−0.03	−0.03	+0.07	−0.25
	−0.40	−0.21	−0.03	+0.15	−0.10	0.00	−0.11	−0.06	−0.04	−0.01	−0.04	−0.11
	−0.35	+0.15	−0.27	+0.25	−0.15	−0.03	−0.08	−0.04	−0.06	−0.00	−0.17	−0.01
	−0.30	+0.28	−0.35	+0.11	−0.03	−0.07	−0.03	−0.01	−0.01	−0.01	−0.21	0.00
	−0.25	+0.11	−0.23	−0.15	+0.13	−0.07	0.00	0.00	+0.01	−0.06	−0.15	−0.08
	−0.20	−0.25	+0.07	−0.35	+0.20	−0.03	0.00	+0.01	+0.03	−0.11	−0.04	−0.10
	−0.15	−0.28	+0.14	−0.28	0.00	+0.03	−0.01	+0.01	0.00	−0.14	+0.01	+0.07
	−0.10	+0.14	+0.03	−0.04	−0.21	+0.07	0.00	+0.01	−0.01	−0.08	0.00	+0.38
	−0.05	+0.76	−0.14	+0.17	−0.37	+0.03	+0.03	+0.01	−0.03	−0.01	−0.03	+0.61
$\Phi_{lr}(\tau)$	0.00	+0.99	−0.17	+0.18	−0.28*	−0.04	+0.04	0.00	−0.06	+0.03	0.00	+0.63
	+0.05	+0.97	−0.04	+0.03	−0.14	−0.14	0.00	−0.04	−0.07	+0.01	+0.11	+0.54
	+0.10	+0.42	+0.39	−0.23	+0.08	−0.20	−0.06	−0.08	−0.04	0.00	+0.34	+0.28
	+0.15	−0.14	+0.79	−0.24	+0.10	−0.17	−0.17	−0.15	−0.06	−0.01	+0.49	+0.04
	+0.20	−0.34	+0.75	+0.07	−0.06	−0.10	−0.23	−0.17	−0.07	+0.08	+0.45	+0.01
	+0.25	−0.03	+0.31	+0.63	−0.27	−0.07	−0.24	−0.20	−0.07	+0.27	+0.24	+0.07
	+0.30	+0.28	−0.11	+0.77	−0.17	−0.10	−0.17	−0.15	0.00	+0.41	+0.06	+0.07
	+0.35	+0.27	−0.14	+0.45	+0.25	−0.13	−0.10	−0.07	+0.11	+0.38	0.00	−0.07
	+0.40	−0.07	+0.03	0.00	+0.76	−0.06	−0.03	+0.03	+0.25	+0.25	+0.06	−0.23
	+0.45	−0.42	+0.18	−0.17	+0.85	+0.13	0.00	+0.11	+0.28	+0.10	+0.08	−0.32
	+0.50	−0.52	+0.14	−0.03	+0.42	+0.32	+0.10	+0.21	+0.27	+0.01	+0.03	−0.25
	+0.55	−0.25	−0.14	+0.25	0.00	+0.46	+0.25	+0.31	+0.20	+0.06	−0.10	−0.08
	+0.60	+0.14	−0.32	+0.28	−0.11	+0.42	+0.44	+0.41	+0.18	+0.11	−0.17	+0.08
	+0.65	+0.31	−0.28	+0.03	+0.23	+0.28	+0.59	+0.46	+0.21	+0.13	−0.11	+0.20
	+0.70	+0.20	−0.04	−0.25	+0.35	+0.21	+0.57	+0.46	+0.24	+0.08	0.00	+0.17
	+0.75	−0.04	+0.20	−0.30	+0.39	+0.25	+0.45	+0.39	+0.21	0.00	+0.11	+0.06
	+0.80	−0.14	+0.27	−0.11	+0.17	+0.28	+0.28	+0.28	+0.14	−0.03	+0.14	−0.04
	+0.85	−0.07	+0.14	+0.15	−0.14	+0.25	+0.14	+0.14	+0.03	−0.03	+0.07	−0.06
	+0.90	+0.14	0.00	+0.28	−0.25	+0.11	+0.10	+0.06	−0.04	+0.03	0.00	+0.01
	+0.95	+0.23	−0.04	+0.20	−0.10	−0.06	+0.07	0.00	−0.04	+0.07	−0.03	+0.03
	+1.00	+0.14	0.00	0.00	+0.08	−0.15	+0.03	−0.01	−0.03	+0.07	0.00	0.00
$\Phi_{ll}(\tau)$	0.00	+1.00	+0.54	+0.39	+0.35	+0.28	+0.34	+0.30	+0.23	+0.27	+0.39	+0.66
$\Phi_{rr}(\tau)$	0.00	+1.00	+1.38	+1.73	+2.06	+1.42	+1.25	+1.13	+0.87	+0.92	+0.75	+0.66

Table D.1 (continued)

	Delay τ [ms]	Horizontal angle ξ 0°	18°	36°	54°	72°	90°	108°	126°	144°	162°	180°
c) Music motif C (Haydn)												
	−1.00	+0.04	−0.10	0.00	−0.02	+0.04	+0.08	+0.04	−0.04	0.00	0.00	−0.02
	−0.95	+0.06	−0.10	−0.02	−0.02	+0.04	+0.08	+0.04	−0.04	+0.02	−0.02	+0.02
	−0.90	+0.08	−0.08	−0.04	−0.02	+0.02	+0.08	+0.02	−0.04	+0.02	−0.02	+0.04
	−0.85	+0.06	−0.04	−0.06	−0.02	+0.02	+0,06	0.00	−0.04	+0.02	−0.02	+0.04
	−0.80	−0.02	0.00	−0.08	−0.04	0.00	+0.04	−0.02	−0.04	0.00	−0.02	+0.06
	−0.75	−0.06	0.00	−0.08	−0.06	−0.02	+0.02	−0.04	−0.04	0.00	−0.02	+0.08
	−0.70	−0.06	−0.02	−0.06	−0.06	−0.04	0.00	−0.04	−0.06	−0.02	−0.02	+0.06
	−0.65	−0.02	−0.06	−0.04	−0.08	−0.08	−0.02	−0.06	−0.06	−0.02	−0.04	+0.04
	−0.60	−0.04	−0.08	−0.06	−0.06	−0.10	−0.04	−0.08	−0.04	−0.04	−0.04	−0.04
	−0.55	−0.14	−0.08	−0.10	−0.04	−0.10	−0.06	−0.10	−0.04	−0.06	−0.04	−0.14
	−0.50	−0.26	−0.04	−0.12	−0.02	−0.10	−0.08	−0.10	−0.02	−0.08	−0.04	−0.20
	−0.45	−0.36	−0.06	−0.10	−0.04	−0.08	−0.10	−0.10	−0.02	−0.08	−0.06	−0.22
	−0.40	−0.30	−0.12	−0.06	−0.06	−0.06	−0.12	−0.08	−0.02	−0.10	−0.10	−0.20
	−0.35	−0.18	−0.22	−0.04	−0.08	−0.06	−0.10	−0.06	0.00	−0.10	−0.14	−0.14
	−0.30	−0.06	−0.30	−0.06	−0.06	−0.04	−0.08	−0.06	0.00	−0.12	−0.18	−0.08
	−0.25	0.00	−0.28	−0.16	−0.04	−0.04	−0.06	−0.04	0.00	−0.14	−0.18	0.00
	−0.20	+0.04	−0.20	−0.24	−0.06	−0.04	−0.06	−0.04	−0.02	−0.16	−0.12	+0.08
	−0.15	+0.18	−0.10	−0.24	−0.12	−0.04	−0.08	−0.06	−0.06	−0.16	−0.06	+0.30
	−0.10	+0.53	0.00	−0.18	−0.22	−0.06	−0.08	−0.08	−0.08	−0.10	0.00	+0.47
	−0.05	+0.79	+0.06	−0.06	−0.30	−0.10	−0.10	−0.12	−0.12	−0.06	+0.10	+0.59
$\Phi_{lr}(\tau)$	0.00	+0.97	+0.18	−0.06	−0.28*	−0.16	−0.14	−0.16	−0.12	−0.04	+0.22	+0.61
	+0.05	+0.97	+0.37	−0.04	−0.26	−0.22	−0.18	−0.22	−0.14	0.00	+0.32	+0.55
	+0.10	+0.65	+0.61	0.00	−0.18	−0.26	−0.24	−0.28	−0.12	+0.06	+0.45	+0.41
	+0.15	+0.36	+0.85	+0.16	−0.12	−0.28	−0.30	−0.30	−0.12	+0.16	+0.55	+0.26
	+0.20	+0.12	+0.83	+0.36	−0.08	−0.26	−0.34	−0.32	−0.08	+0.26	+0.51	+0.14
	+0.25	+0.06	+0.59	+0.59	−0.02	−0.22	−0.36	−0.30	−0.02	+0.36	+0.39	+0.02
	+0.30	+0.02	+0.36	+0.67	+0.14	−0.16	−0.30	−0.20	+0.06	+0.39	+0.30	−0.06

	+0.35	−0.12	+0.20	+0.59	+0.34	−0.06	−0.18	−0.08	+0.18	+0.37	+0.16	−0.18
	+0.40	−0.30	+0.08	+0.39	+0.55	+0.08	−0.04	+0.10	+0.26	+0.32	+0.10	−0.24
	+0.45	−0.41	+0.02	+0.24	+0.61	+0.28	+0.14	+0.28	+0.32	+0.24	+0.04	−0.28
	+0.50	−0.39	−0.06	+0.20	+0.55	+0.43	+0.34	+0.45	+0.34	+0.20	−0.02	−0.24
	+0.55	−0.26	−0.16	+0.18	+0.41	+0.59	+0.53	+0.61	+0.34	+0.16	−0.06	−0.16
	+0.60	−0.10	−0.20	+0.14	+0.34	+0.63	+0.73	+0.73	+0.32	+0.14	−0.06	−0.08
	+0.65	0.00	−0.18	+0.04	+0.32	+0.61	+0.85	+0.79	+0.32	+0.12	−0.04	0.00
	+0.70	0.00	−0.06	−0.04	+0.34	+0.55	+0.85	+0.79	+0.30	+0.10	0.00	+0.04
	+0.75	−0.02	+0.02	−0.04	+0.30	+0.51	+0.77	+0.69	+0.26	+0.10	+0.04	+0.06
	+0.80	−0.02	+0.06	+0.02	+0.20	+0.43	+0.63	+0.55	+0.18	+0.08	+0.04	+0.10
	+0.85	+0.06	+0.06	+0.08	+0.08	+0.32	+0.45	+0.39	+0.10	+0.06	+0.02	+0.10
	+0.90	+0.10	+0.04	+0.10	+0.02	+0.20	+0.30	+0.24	+0.06	+0.04	0.00	+0.10
	+0.95	+0.12	+0.14	+0.06	+0.04	+0.06	+0.20	+0.10	0.00	+0.02	−0.02	+0.08
	+1.00	+0.08	0.00	+0.02	+0.04	−0.04	+0.04	0.00	−0.04	0.00	−0.06	+0.04
$\Phi_{ll}(\tau)$	0.00	+1.00	+0.66	+0.42	+0.40	+0.42	+0.60	+0.57	+0.26	+0.30	+0.45	+0.72
$\Phi_{rr}(\tau)$	0.00	+1.00	+1.17	+1.34	+1.50	+1.39	+1.39	+1.30	+0.87	+0.85	+0.74	+0.58

d) Music motif D (Wagner)

−1.00	−0.10	−0.11	+0.06	−0.04	+0.10	+0.10	+0.08	−0.02	−0.02	+0.08	−0.15
−0.95	+0.06	−0.23	+0.06	−0.04	+0.10	+0.19	+0.11	0.00	0.00	+0.04	−0.06
−0.90	+0.23	−0.23	0.00	+0.02	+0.08	+0.23	+0.15	+0.06	+0.04	0.00	+0.04
−0.85	+0.29	−0.11	−0.10	+0.06	+0.10	+0.19	+0.13	+0.10	+0.02	+0.08	+0.11
−0.80	+0.27	+0.02	−0.15	+0.08	+0.10	+0.13	+0.10	+0.08	−0.04	+0.15	+0.15
−0.75	+0.19	+0.15	−0.13	0.00	+0.10	+0.08	+0.06	+0.04	−0.06	+0.15	+0.21
−0.70	+0.13	+0.19	−0.04	−0.08	+0.06	+0.04	+0.04	−0.02	−0.06	+0.10	+0.27
−0.65	+0.15	+0.15	+0.08	−0.13	0.00	+0.02	0.00	−0.06	−0.04	+0.06	+0.27
−0.60	+0.17	+0.08	+0.10	−0.13	−0.10	−0.02	−0.04	−0.06	+0.02	+0.02	+0.15
−0.55	+0.02	+0.08	+0.06	−0.06	−0.17	−0.10	−0.10	−0.04	+0.06	+0.04	−0.08
−0.50	−0.19	+0.13	0.00	0.00	−0.17	−0.17	−0.13	−0.02	+0.06	+0.10	−0.27
−0.45	−0.36	+0.13	+0.02	+0.06	−0.13	−0.25	−0.17	−0.02	+0.04	+0.10	−0.34
−0.40	−0.34	−0.04	+0.11	+0.06	−0.10	−0.23	−0.13	−0.02	+0.02	0.00	−0.23
−0.35	−0.15	−0.29	+0.17	+0.04	−0.02	−0.15	−0.10	−0.02	0.00	−0.17	−0.11
−0.30	0.00	−0.44	+0.11	+0.04	+0.02	−0.04	−0.04	0.00	−0.02	−0.27	−0.08

Table D.1 (continued)

	Delay τ [ms]	Horizontal angle ξ										
		0°	18°	36°	54°	72°	90°	108°	126°	144°	162°	180°
	−0.25	−0.04	−0.38	−0.11	+0.08	+0.04	+0.06	+0.04	+0.04	−0.06	−0.23	−0.15
	−0.20	−0.13	−0.19	−0.32	+0.10	+0.08	+0.17	+0.10	+0.06	−0.10	−0.11	−0.17
	−0.15	−0.08	−0.06	−0.36	+0.04	+0.11	+0.21	+0.15	+0.04	−0.13	−0.02	+0.06
	−0.10	+0.29	−0.08	−0.15	−0.08	+0.15	+0.23	+0.17	0.00	−0.10	−0.02	+0.50
	−0.05	+0.82	−0.13	0.00	−0.19	+0.19	+0.27	+0.15	−0.04	−0.08	−0.06	+0.82
$\Phi_{lr}(\tau)$	0.00	+1.00	−0.11	+0.02	−0.23*	+0.06	+0.15	+0.10	−0.08	−0.06	−0.06	+0.88
	+0.05	+0.82	+0.15	−0.10	−0.19	−0.08	+0.08	+0.04	−0.10	−0.08	+0.08	+0.72
	+0.10	+0.29	+0.55	−0.21	−0.10	−0.19	−0.04	−0.06	−0.06	−0.08	+0.36	+0.38
	+0.15	−0.08	+0.92	−0.15	−0.08	−0.23	−0.17	−0.13	−0.08	−0.04	+0.55	+0.15
	+0.20	−0.13	+0.76	+0.19	−0.11	−0.21	−0.27	−0.19	−0.11	+0.08	+0.48	+0.11
	+0.25	0.00	+0.25	+0.67	−0.15	−0.21	−0.34	−0.27	−0.13	+0.25	+0.19	+0.06
	+0.30	+0.04	−0.08	+0.76	0.00	−0.23	−0.32	−0.27	−0.06	+0.34	−0.06	−0.76
	+0.35	−0.13	−0.08	+0.40	+0.29	−0.23	−0.27	−0.21	+0.08	+0.31	−0.08	−0.31
	+0.40	−0.38	0.00	+0.04	+0.53	−0.11	−0.19	−0.11	+0.23	+0.21	0.00	−0.46
	+0.45	−0.46	−0.06	−0.10	+0.51	+0.10	−0.10	+0.02	+0.27	+0.06	+0.04	−0.40
	+0.50	−0.32	−0.19	+0.04	+0.31	+0.36	+0.04	+0.17	+0.25	−0.02	−0.04	−0.19
	+0.55	−0.10	−0.38	+0.13	+0.04	+0.48	+0.25	+0.31	+0.15	0.00	−0.17	−0.02
	+0.60	+0.10	−0.40	+0.04	0.00	+0.42	+0.48	+0.40	+0.13	+0.04	−0.23	+0.04
	+0.65	+0.13	−0.23	−0.21	+0.06	+0.29	+0.61	+0.46	+0.19	+0.06	−0.15	+0.04
	+0.70	+0.13	0.00	−0.32	+0.11	+0.27	+0.61	+0.46	+0.23	+0.04	+0.02	+0.11
	+0.75	+0.19	+0.15	−0.23	+0.08	+0.29	+0.44	+0.38	+0.19	+0.02	+0.11	+0.23
	+0.80	+0.23	+0.15	−0.04	−0.06	+0.27	+0.29	+0.25	+0.08	−0.02	+0.06	+0.31
	+0.85	+0.25	+0.13	+0.15	−0.17	+0.15	+0.10	+0.08	−0.06	−0.02	−0.06	+0.29
	+0.90	+0.17	+0.17	+0.15	−0.13	−0.02	+0.04	−0.04	−0.10	−0.02	−0.15	+0.17
	+0.95	0.00	+0.23	+0.06	−0.04	−0.21	−0.06	−0.10	−0.10	+0.02	−0.15	0.00
	+1.00	−0.10	+0.23	+0.02	+0.04	−0.27	−0.15	−0.13	−0.06	+0.04	−0.08	−0.15
$\Phi_{ll}(\tau)$	0.00	+1.00	+0.65	+0.46	+0.20	+0.28	+0.36	+0.28	+0.22	+0.23	+0.44	+0.90
$\Phi_{rr}(\tau)$	0.00	+1.00	+1.40	+1.66	+1.66	+1.56	+1.38	+1.14	+0.97	+0.87	+0.80	+0.96

e) Music motif E (Mozart)

$\Phi_{lr}(\tau)$											
−1.00	−0.07	−0.06	+0.03	−0.16	−0.07	−0.01	−0.03	−0.05	−0.07	−0.01	−0.08
−0.95	−0.09	−0.01	0.00	−0.03	−0.10	0.00	−0.05	−0.04	−0.02	±0.00	−0.09
−0.90	−0.10	−0.02	−0.04	+0.05	−0.11	−0.04	−0.07	−0.04	−0.03	−0.01	−0.10
−0.85	−0.08	−0.09	−0.06	+0.03	−0.08	−0.10	−0.07	−0.04	−0.04	−0.03	−0.10
−0.80	−0.06	−0.12	−0.04	−0.05	−0.05	−0.08	−0.04	−0.04	−0.05	−0.07	−0.08
−0.75	−0.06	−0.10	−0.04	−0.06	−0.04	−0.05	−0.04	−0.04	−0.06	−0.13	−0.07
−0.70	−0.08	−0.08	−0.05	−0.05	−0.03	−0.04	−0.03	−0.04	−0.08	−0.15	−0.10
−0.65	−0.12	−0.08	−0.09	−0.03	−0.04	0.00	−0.03	−0.07	−0.10	−0.14	−0.15
−0.60	−0.12	−0.12	−0.13	−0.05	−0.06	−0.01	−0.01	−0.08	−0.12	−0.13	−0.18
−0.55	−0.12	−0.17	−0.14	−0.08	−0.04	−0.03	−0.03	−0.09	−0.12	−0.15	−0.19
−0.50	−0.14	−0.19	−0.13	−0.12	−0.07	−0.05	−0.05	−0.13	−0.12	−0.17	−0.22
−0.45	−0.20	−0.17	−0.13	−0.16	−0.11	−0.08	−0.07	−0.15	−0.12	−0.18	−0.27
−0.40	−0.29	−0.18	−0.15	−0.16	−0.15	−0.11	−0.11	−0.16	−0.13	−0.21	−0.32
−0.35	−0.29	−0.25	−0.16	−0.15	−0.19	−0.12	−0.12	−0.15	−0.15	−0.27	−0.33
−0.30	−0.17	−0.36	−0.17	−0.16	−0.19	−0.14	−0.12	−0.16	−0.17	−0.34	−0.29
−0.25	−0.03	−0.37	−0.18	−0.17	−0.19	−0.16	−0.15	−0.22	−0.21	−0.38	−0.21
−0.20	+0.04	−0.26	−0.21	−0.21	−0.19	−0.16	−0.19	−0.27	−0.26	−0.40	−0.08
−0.15	+0.10	−0.09	−0.28	−0.30	−0.23	−0.17	−0.25	−0.33	−0.32	−0.32	+0.13
−0.10	+0.33	−0.04	−0.31	−0.38	−0.30	−0.21	−0.27	−0.38	−0.34	−0.22	+0.38
−0.05	+0.74	−0.03	−0.29	−0.41	−0.37	−0.27	−0.31	−0.43	−0.31	−0.06	+0.71
0.00	+0.98	+0.21	−0.19	−0.38	−0.42*	−0.34	−0.37	−0.46	−0.26	+0.18	+0.90
+0.05	+0.68	+0.57	−0.08	−0.40	−0.41	−0.38	−0.39	−0.42	−0.14	+0.43	+0.72
+0.10	+0.18	+0.81	−0.07	−0.29	−0.43	−0.43	−0.46	−0.40	−0.01	+0.63	+0.41
+0.15	−0.04	+0.63	0.00	−0.24	−0.45	−0.49	−0.50	−0.37	+0.15	+0.69	+0.14
+0.20	−0.05	+0.29	+0.21	−0.24	−0.45	−0.50	−0.49	−0.23	+0.31	+0.57	−0.01
+0.25	−0.08	+0.07	+0.47	−0.19	−0.41	−0.49	−0.46	−0.05	+0.43	+0.32	−0.11
+0.30	−0.21	+0.02	+0.56	+0.03	−0.31	−0.43	−0.34	+0.13	+0.45	+0.13	−0.19
+0.35	−0.26	−0.03	+0.41	+0.38	−0.19	−0.34	−0.19	+0.31	+0.36	−0.02	−0.21
+0.40	−0.25	−0.09	+0.19	+0.69	0.00	−0.16	−0.01	+0.46	+0.21	−0.07	−0.21
+0.45	−0.16	−0.13	+0.05	+0.75	+0.25	+0,04	+0.20	+0.53	+0.09	−0.09	−0.18
+0.50	−0.08	−0.13	0.00	+0.53	+0.52	+0.31	+0.46	+0.48	+0.02	−0.09	−0.14
+0.55	−0.06	−0.08	−0.01	+0.28	+0.79	+0.64	+0.69	+0.41	+0.01	−0.07	−0.10

Table D.1 (continued)

	Delay τ [ms]	Horizontal angle ξ										
		0°	18°	36°	54°	72°	90°	108°	126°	144°	162°	180°
	+0.60	−0.09	−0.01	−0.01	+0.20	+0.79	+0.89	+0.79	+0.37	+0.03	−0.04	−0.07
	+0.65	−0.09	+0.01	−0.01	+0.23	+0.69	+1.07	+0.80	+0.35	+0.05	−0.01	−0.06
	+0.70	−0.07	0.00	0.00	+0.27	+0.61	+1.05	+0.80	+0.37	+0.07	+0.03	−0.04
	+0.75	−0.06	−0.02	+0.03	+0.30	+0.50	+0.79	+0.73	+0.37	+0.09	+0.04	−0.03
	+0.80	−0.05	0.00	+0.05	+0.25	+0.40	+0.46	+0.54	+0.30	+0.09	+0.05	−0.03
	+0.85	−0.07	−0.01	+0.03	+0.16	+0.25	+0.22	+0.28	+0.16	+0.07	+0.04	−0.05
	+0.90	−0.09	−0.04	0.00	+0.08	+0.08	+0.08	+0.06	+0.05	+0.02	−0.01	−0.09
	+0.95	−0.09	−0.09	+0.05	+0.01	+0.05	−0.03	−0.03	0.00	−0.03	−0.06	−0.10
	+1.00	−0.05	−0.13	+0.07	−0.07	+0.11	−0.07	−0.04	−0.08	−0.07	−0.10	−0.09
$\Phi_{ll}(\tau)$	0.00	+1.00	+0.73	+0.31	+0.39	+0.44	+0.60	+0.47	+0.35	+0.28	+0.63	+0.87
$\Phi_{rr}(\tau)$	0.00	+1.00	+1.31	+1.55	+2.61	+2.51	+2.44	+2.30	+1.90	+1.28	+1.13	+0.97
f) Continuous speech (female)												
	−1.00	−0.18	−0.21	−0.12	−0.12	−0.18	−0.18	−0.26	−0.15	−0.12	−0.15	−0.18
	−0.95	−0.15	−0.24	−0.15	−0.09	−0.15	−0.12	−0.21	−0.12	−0.15	−0.18	−0.15
	−0.90	−0.12	−0.24	−0.18	−0.09	−0.09	−0.06	−0.15	−0.09	−0.15	−0.18	−0.10
	−0.85	−0.06	−0.24	−0.21	−0.12	−0.06	0.00	−0.09	−0.09	−0.15	−0.15	−0.04
	−0.80	0.00	−0.18	−0.24	−0.12	−0.06	+0.03	−0.06	−0.09	−0.15	−0.12	+0.03
	−0.75	0.00	−0.15	−0.24	−0.15	−0.06	+0.06	−0.03	−0.09	−0.15	−0.09	+0.08
	−0.70	+0.03	−0.09	−0.24	−0.18	−0.06	+0.03	−0.06	−0.09	−0.15	−0.06	+0.12
	−0.65	0.00	−0.03	−0.21	−0.21	−0.09	0.00	−0.09	−0.12	−0.15	−0.03	+0.15
	−0.60	0.00	0.00	−0.18	−0.21	−0.12	−0.06	−0.12	−0.12	−0.15	0.00	+0.16
	−0.55	−0.06	0.00	−0.15	−0.21	−0.18	−0.12	−0.18	−0.15	−0.15	0.00	+0.15
	−0.50	−0.09	0.00	−0.09	−0.21	−0.21	−0.18	−0.21	−0.15	−0.12	0.00	+0.14
	−0.45	−0.09	−0.03	−0.06	−0.18	−0.24	−0.24	−0.24	−0.15	−0.09	0.00	+0.14
	−0.40	−0.06	−0.06	−0.09	−0.15	−0.24	−0.26	−0.24	−0.15	−0.09	−0.03	+0.15
	−0.35	+0.03	−0.09	−0.09	−0.15	−0.21	−0.29	−0.24	−0.12	−0.09	−0.06	+0.20

	−0.30	+0.15	−0.09	−0.09	−0.12	−0.18	−0.29	−0.21	−0.09	−0.09	−0.06	+0.27
	−0.25	+0.32	−0.06	−0.12	−0.12	−0.15	−0.26	−0.15	−0.09	−0.09	−0.03	+0.38
	−0.20	+0.56	+0.03	−0.15	−0.15	−0.15	−0.21	−0.12	−0.09	−0.09	+0.03	+0.48
	−0.15	+0.71	+0.15	−0.12	−0.15	−0.12	−0.18	−0.06	−0.09	−0.06	+0.09	+0.59
	−0.10	+0.88	+0.32	−0.06	−0.15	−0.12	−0.15	−0.03	−0.06	−0.03	+0.24	+0.71
	−0.05	+0.97	+0.50	+0.06	−0.15	−0.12	−0.15	−0.06	−0.06	+0.03	+0.38	+0.75
$\Phi_{lr}(\tau)$	0.00	+1.00	+0.59	+0.12	−0.15	−0.15	−0.18*	−0.12	−0.09	+0.06	+0.44	+0.76
	+0.05	+0.97	+0.74	+0.24	−0.15	−0.18	−0.18	−0.15	−0.12	+0.09	+0.59	+0.72
	+0.10	+0.88	+0.85	+0.38	−0.06	−0.15	−0.21	−0.15	−0.06	+0.24	+0.68	+0.66
	+0.15	+0.74	+0.91	+0.56	+0.06	−0.15	−0.24	−0.15	0.00	+0.35	+0.74	+0.57
	+0.20	+0.56	+0.88	+0.71	+0.24	−0.09	−0.24	−0.12	+0.09	+0.50	+0.76	+0.45
	+0.25	+0.35	+0.82	+0.82	+0.41	0.00	−0.18	−0.06	+0.21	+0.59	+0.74	+0.34
	+0.30	+0.21	+0.68	+0.88	+0.59	+0.12	−0.09	+0.06	+0.35	+0.65	+0.68	+0.24
	+0.35	+0.06	+0.53	+0.85	+0.71	+0.26	+0.06	+0.18	+0.47	+0.68	+0.56	+0.15
	+0.40	0.00	+0.38	+0.79	+0.82	+0.41	+0.21	+0.35	+0.59	+0.71	+0.44	+0.12
	+0.45	−0.06	+0.24	+0.71	+0.88	+0.59	+0.41	+0.53	+0.68	+0.68	+0.35	+0.09
	+0.50	−0.03	+0.15	+0.59	+0.88	+0.71	+0.62	+0.71	+0.74	+0.65	+0.29	+0.09
	+0.55	0.00	+0.09	+0.47	+0.82	+0.82	+0.79	+0.88	+0.76	+0.59	+0.24	+0.10
	+0.60	+0.03	+0.09	+0.35	+0.76	+0.91	+0.94	+1.00	+0.76	+0.53	+0.21	+0.12
	+0.65	+0.06	+0.12	+0.29	+0.68	+0.91	+1.03	+1.06	+0.74	+0.47	+0.21	+0.12
	+0.70	+0.06	+0.15	+0.26	+0.62	+0.88	+1.06	+1.06	+0.71	+0.41	+0.24	+0.12
	+0.75	+0.06	+0.18	+0.26	+0.56	+0.82	+1.03	+1.00	+0.62	+0.38	+0.24	+0.09
	+0.80	+0.03	+0.21	+0.26	+0.47	+0.74	+0.91	+0.88	+0.53	+0.35	+0.21	+0.04
	+0.85	−0.03	+0.21	+0.29	+0.44	+0.62	+0.79	+0.74	+0.44	+0.29	+0.18	−0.01
	+0.90	−0.09	+0.15	+0.29	+0.41	+0.50	+0.62	+0.59	+0.35	+0.26	+0.15	−0.07
	+0.95	−0.15	+0.09	+0.26	+0.35	+0.41	+0.47	+0.44	+0.26	+0.21	+0.06	−0.12
	+1.00	−0.18	0.00	+0.21	+0.32	+0.29	+0.32	+0.32	+0.24	+0.15	0.00	−0.15
$\Phi_{ll}(\tau)$	0.00	+1.00	+0.82	+0.68	+0.65	+0.76	+0.71	+0.76	+0.56	+0.56	+0.68	+0.81
$\Phi_{rr}(\tau)$	0.00	+1.00	+1.32	+1.68	+1.85	+1.76	+1.71	+1.76	+1.79	+1.35	+1.06	+0.81

Table D.1 (continued)

	Delay τ [ms]	Horizontal angle ξ 0°	18°	36°	54°	72°	90°	108°	126°	144°	162°	180°
g) Filtered Gaussian noise, 0.25...2.0 kHz[a]												
	−1.00	−0.52	−0.25	+0.12	+0.19	+0.02	−0.12	−0.14	+0.06	+0.06	−0.21	−0.50
	−0.95	−0.41	−0.37	0.00	+0.17	+0.06	−0.06	0.00	+0.12	+0.02	−0.31	−0.43
	−0.90	−0.21	−0.45	−0.12	+0.12	+0.12	+0.04	+0.12	+0.14	−0.04	−0.35	−0.27
	−0.85	+0.06	−0.48	−0.27	+0.06	+0.14	+0.14	+0.25	+0.12	−0.10	−0.35	−0.06
	−0.80	+0.31	−0.35	−0.35	−0.04	+0.14	+0.23	+0.33	+0.10	−0.17	−0.27	+0.21
	−0.75	+0.56	−0.17	−0.37	−0.14	+0.12	+0.27	+0.35	+0.04	−0.19	−0.10	+0.43
	−0.70	+0.64	+0.06	−0.31	−0.23	+0.06	+0.25	+0.31	−0.02	−0.19	+0.06	+0.58
	−0.65	+0.62	+0.33	−0.21	−0.27	−0.02	+0.19	+0.19	−0.10	−0.14	+0.25	+0.58
	−0.60	+0.43	+0.54	−0.04	−0.27	−0.08	+0.06	0.00	−0.17	−0.06	+0.39	+0.48
	−0.55	+0.12	+0.66	+0.17	−0.21	−0.17	−0.08	−0.21	−0.19	+0.04	+0.45	+0.25
	−0.50	−0.21	+0.62	+0.33	−0.10	−0.21	−0.23	−0.35	−0.21	+0.14	+0.43	−0.04
	−0.45	−0.52	+0.45	+0.45	+0.04	−0.23	−0.35	−0.48	−0.17	+0.23	+0.31	−0.33
	−0.40	−0.76	+0.17	+0.48	+0.19	−0.19	−0.41	−0.52	−0.08	+0.27	+0.10	−0.50
	−0.35	−0.83	−0.25	+0.37	+0.31	−0.10	−0.39	−0.41	+0.02	+0.25	−0.19	−0.74
	−0.30	−0.76	−0.58	+0.17	+0.37	+0.02	−0.27	−0.25	+0.10	+0.14	−0.41	−0.72
	−0.25	−0.48	−0.79	−0.06	+0.33	+0.14	−0.08	0.00	+0.19	+0.02	−0.58	−0.43
	−0.20	−0.04	−0.87	−0.33	+0.21	+0.25	+0.12	+0.27	+0.25	−0.12	−0.62	−0.25
	−0.15	+0.37	−0.79	−0.54	+0.04	+0.31	+0.33	+0.50	+0.23	−0.27	−0.56	+0.14
	−0.10	+0.68	−0.48	−0.66	−0.17	+0.31	+0.50	+0.62	+0.17	−0.35	−0.33	+0.48
	−0.05	+0.85	−0.04	−0.62	−0.35	+0.23	+0.54	+0.62	+0.04	−0.35	−0.04	+0.68
$\Phi_{lr}(\tau)$	0.00	+0.95	+0.37	−0.50*	−0.50*	+0.10	+0.50	+0.50	−0.06	−0.27	+0.25	+0.79
	+0.05	+0.85	+0.74	−0.27	−0.52	−0.04	+0.31	+0.17	−0.19	−0.12	+0.54	+0.70
	+0.10	+0.68	+0.95	+0.06	−0.43	−0.21	+0.04	−0.21	−0.27	+0.06	+0.72	+0.48
	+0.15	+0.37	+0.97	+0.41	−0.23	−0.35	−0.27	−0.50	−0.33	+0.21	+0.74	+0.17
	+0.20	−0.04	+0.74	+0.64	0.00	−0.41	−0.52	−0.76	−0.31	+0.33	+0.58	−0.21
	+0.25	−0.48	+0.35	+0.72	+0.23	−0.41	−0.70	−0.87	−0.23	+0.37	+0.31	−0.52
	+0.30	−0.76	−0.12	+0.64	+0.43	−0.31	−0.72	−0.79	−0.12	+0.31	−0.06	−0.68

	+0.35	−0.83	−0.56	+0.48	+0.52	−0.17	−0.62	−0.54	+0.04	+0.19	−0.37	−0.72
	+0.40	−0.76	−0.89	+0.12	+0.50	+0.06	−0.31	−0.12	+0.19	0.00	−0.62	−0.60
	+0.45	−0.52	−0.93	−0.21	+0.33	+0.25	0.00	+0.25	+0.29	−0.14	−0.70	−0.33
	+0.50	−0.21	−0.81	−0.52	+0.12	+0.37	+0.37	+0.62	+0.33	−0.27	−0.62	−0.10
	+0.55	+0.12	−0.56	−0.66	−0.10	+0.45	+0.62	+0.87	+0.31	−0.31	−0.41	+0.25
	+0.60	+0.43	−0.21	−0.66	−0.33	+0.41	+0.79	+0.93	+0.23	−0.29	−0.14	+0.45
	+0.65	+0.62	+0.21	−0.54	−0.45	+0.31	+0.76	+0.79	+0.10	−0.21	+0.17	+0.58
	+0.70	+0.64	+0.54	−0.31	−0.45	+0.14	+0.60	+0.50	−0.04	−0.08	+0.43	+0.56
	+0.75	+0.56	+0.70	0.00	−0.39	−0.06	+0.29	+0.12	−0.19	+0.04	+0.54	+0.43
	+0.80	+0.31	+0.74	+0.27	−0.23	−0.21	−0.04	−0.31	−0.29	+0.14	+0.56	+0.21
	+0.85	+0.06	+0.60	+0.45	−0.04	−0.31	−0.41	−0.66	−0.33	+0.19	+0.43	−0.04
	+0.90	−0.21	+0.33	+0.54	+0.14	−0.39	−0.64	−0.83	−0.29	+0.21	+0.23	−0.27
	+0.95	−0.41	0.00	+0.52	+0.27	−0.37	−0.74	−0.85	−0.21	+0.14	0.00	−0.41
	+1.00	−0.52	−0.27	+0.37	+0.33	−0.29	−0.68	−0.68	−0.10	+0.08	−0.21	−0.50
$\Phi_{ll}(\tau)$	0.00	+1.00	+0.75	+0.45	+0.19	+0.11	+0.38	+0.45	+0.11	+0.16	+0.53	+0.82
$\Phi_{rr}(\tau)$	0.00	+1.00	+1.35	+1.55	+1.71	+1.86	+1.79	+1.57	+1.37	+1.12	+1.10	+0.84

[a] In the calculations published in [3.3], the correlation values listed here were used

Table D.2. Average values of correlation functions at $\tau = 0°$ for the five music motifs (A, B, C, D, and E) as a function of the horizontal angle of incidence ξ ($\eta = 0°$). These values may be used to calculate the IACC, when designing concert halls. For arbitrary angles of incidence ξ, η the angle given by $\sin^{-1}(\sin\xi \cos\eta)$ is substituted for the horizontal angle ξ

	Horizontal angle ξ										
	0°	18°	36°	54°	72°	90°	108°	126°	144°	162°	180°
$\Phi_{lr}(0)$	0.99	0.08	−0.07	−0.30*	−0.09	−0.03	−0.09	−0.16	−0.08	0.13	0.74
$\Phi_{ll}(0)$	1.00	0.66	0.40	0.33	0.35	0.51	0.45	0.25	0.26	0.49	0.77
$\Phi_{rr}(0)$	1.00	1.28	1.52	1.85	1.63	1.59	1.48	1.07	0.95	0.83	0.78

Table D.3. Measured correlation functions at $\tau = 0$ for each 1/3 octave bandpass noise as a function of the horizontal angle of incidence ξ ($\eta = 0°$). For the range $180° < \xi < 0°$, the values are obtained by setting $\xi = 360° - \xi$ and interchanging the suffixes l and r

Center frequency		Horizontal angle ξ										
		0°	18°	36°	54°	72°	90°	108°	126°	144°	162°	180°
125	$\Phi_{lr}(0)$	1.00	0.99	0.95	0.93	0.91	0.89*	0.90	0.92	0.93	0.99	1.02
	$\Phi_{ll}(0)$	1.00	0.98	1.02	1.00	1.00	1.02	1.00	0.95	0.96	1.03	1.02
	$\Phi_{rr}(0)$	1.00	1.01	1.05	1.10	1.08	1.16	1.08	1.04	1.05	1.04	1.00
250	$\Phi_{lr}(0)$	0.98	0.95	0.87	0.62	0.40	0.31*	0.38	0.55	0.78	0.98	1.03
	$\Phi_{ll}(0)$	1.00	1.02	1.16	1.17	1.15	1.14	1.10	1.14	1.11	1.10	1.05
	$\Phi_{rr}(0)$	1.00	1.11	1.26	1.29	1.29	1.32	1.32	1.29	1.26	1.16	1.04
500	$\Phi_{lr}(0)$	1.00	0.84	0.13	−0.54	−1.05	−1.25*	−1.10	−0.68	−0.04	0.58	0.94
	$\Phi_{ll}(0)$	1.00	0.87	0.87	1.00	1.05	1.11	0.98	0.95	0.78	0.71	0.90
	$\Phi_{rr}(0)$	1.00	1.39	1.61	2.36	2.52	2.65	2.59	2.52	1.76	1.29	1.03
1000	$\Phi_{lr}(0)$	1.00	0.61	−0.29	−0.72*	−0.65	−0.58	−0.64	−0.69	−0.33	0.39	0.87
	$\Phi_{ll}(0)$	1.00	0.51	0.25	0.26	0.49	0.63	0.51	0.25	0.18	0.41	0.88
	$\Phi_{rr}(0)$	1.00	1.35	1.60	2.10	2.45	2.54	2.51	2.31	1.91	1.42	0.92
2000	$\Phi_{lr}(0)$	0.99	−0.28	−0.64*	0.19	0.06	−0.24	−0.05	0.02	−0.58	−0.31	0.97
	$\Phi_{ll}(0)$	1.00	0.62	0.51	0.16	0.10	0.37	0.14	0.24	0.40	0.51	1.00
	$\Phi_{rr}(0)$	1.00	1.40	1.59	1.53	1.58	1.53	1.39	1.34	1.26	1.23	0.98
4000	$\Phi_{lr}(0)$	1.00	−0.66*	0.20	−0.08	−0.02	−0.09	0.03	0.01	0.05	−0.08	0.11
	$\Phi_{ll}(0)$	1.00	0.36	0.10	0.10	0.06	0.04	0.09	0.06	0.10	0.10	0.11
	$\Phi_{rr}(0)$	1.00	1.57	1.64	1.65	1.47	0.96	0.35	0.10	0.10	0.11	0.11

* The most effective horizontal angle to decrease the IACC.

The magnitude of the IACC is simply obtained at the time of origin which usually has the maximum value. It is remarkable that the calculated values of the IACC scarcely differ between source signals because it is mainly characterized by sound transmission around head and pinnae, so that it is called the spatial-binaural criterion.

Average correlation values for five music motifs are listed in Table D.2, to calculate the IACC of concert halls. These values are also plotted in Fig. 3.2. Table D.3 lists measured values of correlations of each 1/3 octave bandpass filtered noise for any sound source signals with known spectrum.

E. Computation Programs for the Fast Method of Measuring Impulse Responses

(Computer: Univac 1100/83; ASC11 FORTRAN Compiler, ANSI STANDARD X3.9 – 1978)

```
c       demonstration program:
c       mls deconvolution via fast hadamard transform
c       H. Alrutz, 1/10/82
c       Drittes Physikalisches Institut
c       Universitaet Goettingen, W. Germany
c
c
        implicit integer (a-z)
        dimension s(1024),d(512),p(512),h(10)
        dimension taps(20),tob(20)
        data h/5,0,0,3,0,2,0,0,0,1/
c
        read(5,*) ntap,(taps(k),k=1,ntap)
        read(5,*) ntob,(tob(k),k=1,ntob)
c
        call gprm(ntap,taps,ntob,tob,d,p,512)
c
        m=taps(ntap)
        n=2**m
c
c       generate test sequence (+1/-1)
        do 10 k=1,n-1
        s(k)=(-1)**d(k)
        s(k+n-1)=s(k)
10      continue
c
c       convolute periodic seq. and imp. res.
        d(0)=0
        do 20 k=2,n
        d(k)=0
        do 20 j=1,10
        d(k)=d(k)+h(j)*s(n-1+k-j)
20      continue
c
        print 1020,(p(k),k=1,n)
1020    format(//,' permutation ',//,(8i8))
        print 1030,(d(k),k=2,n)
1030    format(//,' input sequence ',//,(8i8))
```

```
c       permut data & permutation
c
        do 40 l=2,n
        if (p(l).le.0) goto 40
        i=0
        j=l
        a=d(j)
30      k=p(j)
        p(j)=-i
        if (k.le.0) goto 40
        b=d(k)
        d(k)=a
        a=b
        i=j
        j=k
        goto 30
40      continue
c
c       fast hadamard transform
c
        ni=1
        do 50 i=1,m
        nj=ni
        ni=ni*2
        do 60 k=0,n-1,ni
        do 60 j=1,nj
        a=d(k+j)
        b=d(k+j+nj)
        d(k+j)=a+b
        d(k+j+nj)=a-b
60      continue
50      continue
c
c       permut data & permutation
c
        do 140 l=2,n
        if (p(l).gt.0) goto 140
        i=0
        j=l
        a=d(j)
130     k=-p(j)
        p(j)=i
        if (k.le.0) goto 140
        b=d(k)
        d(k)=a-d(1)
        a=b
        i=j
        j=k
        goto 130
140     continue
```

```
c
c       data are in time reversed order
c
        print 1040,d(2),(d(k),k=n,3,-1)
1040    format(//,' result ',//,(8i8))
        stop
        end
```

```
        subroutine gprm(ntap,taps,gtob,tob,mxf,permut,nm)
c       generate permutation for mxf-hadamard transform
c
c       H. Alrutz 1/10/82
c
c       ntap            number of sr-taps
c       taps            sr-taps (in ascending order)
c       gtob = 0        generate tob
c       tob             trace orthogonal base
c       mxf             scratch array
c       permut          permutation
c
        implicit integer (a-z)
        dimension taps(1),tob(1),mxf(0:0),permut(1)
c
c
        n=taps(ntap)
        p=2**n-1
        if (p.gt.nm) then
        print *,' sr-length gt ',nm
        stop
        end if
c
c       generate idempotent shift of ml-sequence
c
        s=n
        do 10 i=0,p
        do 20 k=1,ntap
        if (i-taps(k)) 50,30,40
30      s=s+i
        goto 50
40      s=s+mxf(i-taps(k))
20      continue
50      mxf(i)=mod(s,2)
        s=0
10      continue
```

```
c
      if (gtob.eq.0) then
c
c     look up TOB
c
      j=0
      i=0
      goto 120
110   i=i+1
      if (i.lt.p) goto 120
      i=tob(j)
      j=j-1
      if (j.ge.0) goto 110
      print *,' no TOB found, (wrong taps?) '
      stop
c
120   if (mxf(i).ne.1) goto 110
      do 130 k=1,j
      if (mxf(mod(i+tob(k),p)).eq.1) goto 110
130   continue
c
      j=j+1
      tob(j)=i
      if (j.ne.n) goto 110
c
      print 1020,n,(taps(k),k=1,ntap)
1020  format(' sr-length',i6,/,' taps ',(10i6))
      print 1030,(tob(k),k=1,n)
1030  format(' TOB  ',(10i6))
      end if
c
c
c     generate permutation for fast hadamard transform
c
      do 210 i=0,p-1
      s=0
      do 220 k=1,n
      s=s*2+mxf(mod(tob(k)+i,p))
220   continue
      permut(i+2)=s+1
210   continue
      permut(1)=1
c
      return
      end
```

sr-length	**6**					
taps	**1**	**6**				
TOB	**1**	**6**	**8**	**29**	**43**	**48**

permutation

1	64	33	57	39	62	48	17
49	9	47	20	61	45	29	21
59	42	22	58	38	50	12	35
56	15	44	27	18	52	5	11
34	60	43	26	30	24	55	14
40	63	36	53	3	16	41	23
54	2	4	13	37	51	8	7
6	10	46	32	25	31	28	19

input sequence

7	1	1	1	-7	-5	-9	1
-9	1	-5	-1	-1	-11	9	-1
-5	-5	7	-7	-5	-5	-1	-5
-9	5	-11	5	5	-11	9	7
-7	-1	-3	3	5	1	-5	7
1	-7	-1	-11	5	1	-9	5
-3	3	9	-1	-1	-3	11	3
5	5	-3	11	11	3	9	

result

320	0	0	192	0	128	0	0
0	64	0	0	0	0	0	0
0	0	0	0	0	0	0	0
0	0	0	0	0	0	0	0
0	0	0	0	0	0	0	0
0	0	0	0	0	0	0	0
0	0	0	0	0	0	0	0
0	0	0	0	0	0	0	

Glossary of Symbols

A small letter, for example, $p(t)$, is used for real functions in the time domain, while its Fourier transform is represented by a capital letter $P(\omega)$ in the complex frequency domain. The symbol $|\,|$ means the amplitude of the absolute value and $\langle\ \rangle$ means the ensemble. An asterisk signifies the complex conjugate or convolution. The symbol $\rightarrow$ means "approaches"; the symbol $\approx$ means "approximately equal to".

In the following list the number in parentheses indicates the equation where the symbol is defined or used.

a	Plate dimension (2.11)
A	Total amplitude of reflections (5.6)
A_{11}	Amplitude of auditory evoked potential
a_i	Boundaries of ith wells of a diffusing wall
a_n	Lower boundary of nth rectangular region in periodic structure
A_n	Pressure amplitude determined by the $(1/r)$ law, A_0 being unity (3.2)
$A_{n,m}$	Field amplitudes (2.27)
b	Plate dimension (2.11)
b_n	Upper boundary of nth rectangular region in periodic structure
$B_{n,m}$	Field amplitudes (2.27)
c	Speed of sound in air
c'	Constant to be determined (C.1, 2)
$C(K)$	Real part of Fresnel integral (2.18)
C_m	Fourier transform of the exponential sequence c_n (6.9)
c_n	$= \exp(\pm j2\pi s_n/N)$ (6.8)
$C(x)$	Membrane compliance (2.51)
$C(\omega)$	Transfer function of the middle ear (2.48)
d_n	Depth of each well based on the quadratic-residue sequence (6.7)
$d(t)$	Sound source as a disturbance in a hall when measuring impulse response (7.7)
$d(x)$	Depth of each well of diffusing ceiling (6.6)
$D(\Delta t_n)$	Energy ratio between early reflections and early-plus-subsequent reverberation. When $\Delta t_n = 50$ ms it becomes the "Deutlichkeit", and $D(\Delta t_n) = 1/(1+A^2)$, $\Delta t_n \rightarrow 0$ (7.27)
e	$= 2.71828\ldots$

E_K Equilibrium potential of K^+ ion (2.55)
EPSP Excitatory postsynaptic potential
E_{Na} Equilibrium potential of Na^+ ion (2.56)
$E(\omega)$ Transfer function of the external canal (2.48)
$f = \omega/2\pi$ Frequency
F Faraday's constant (2.55, 56)
$f_{l,r}(t)$ Sound pressures at both ear-canal entrances (3.1)
$f_{\pm r}$ Critical frequencies (2.41)
g Gain of comb filter (3.15); and of all-pass filter (3.23)
$g(t)$ Impulse response between disturbing source and observation (7.7)
$g(x)$ Mapping function from the physically realizable space to the auditory subjective space (1.1, 5.9)
g^{-1} Inverse function from the auditory subjective space to the physically realizable space
H Hadamard matrix of Silvester-type (7.21)
$h_{l,r}(r|r_0; t)$ Impulse responses between a sound source located at r_0 and the left and right ear-canal entrances of a listener sitting at r (3.2)
$h_{l,r1}(t)$ Impulse responses for paths from two loudspeakers L_1 and L_2 to
$h_{l,r2}(t)$ entrances of left and right ear canals, respectively (3.9)
h_n Depth on nth rectangular region in a periodic structure
$h_{nl,r}(t)$ Impulse responses from a free field to both ear-canal entrances (3.2)
$h(t)$ Impulse response of a system (3.15, 7.4, 12)
$H(\omega)$ Transfer function
i Integer
IACC $= |\phi_{lr}(\tau)|_{max}$ for $|\tau| \leqq 1$ ms (3.8)
IPSP Inhibitory postsynaptic potential
$j = \sqrt{-1} = -i$ Imaginary unit
j Integer (7.28 – 30)
$k = \omega/c = 2\pi/\lambda$ Wave number
$\bar{k}$ Magnitude of coherence between input signals to loudspeakers located around a listener (Appendix A)
k' Constant (C.1, 2)
$[K^+]$ Concentration of K^+ ion (2.55)
l Distance (2.12); period of structure of walls
L Relative sound-pressure level (5.20)
$L(x)$ Membrane mass (2.51)
m Integer; distance (2.12)
$m(x)$ Mass of each section of basilar membrane (2.49)
n Integer; single sound with a horizontal angle ξ and an elevation angle η to a listener

$[Na^+]$ Concentration of Na^+ ions (2.56)
N Period (6.5); Constant power spectrum (7.5)
$n(t)$ Stationary white noise (7.22)
$P(B>A)$ probability that B is preferred to A (4.4)
$P_d(\omega)$ Power density spectrum (2.4)
P_i, P_r Incident and reflection powers across one period and per unit length in the z direction, respectively (2.39)
$p_n(t)$ Radiation pattern of sound source for each sound n
$p(t)$ Source signal (2.1, 3.1)
PWL Power level of sound source (5.20)
$P(x, \omega)$ Half the pressure across the basilar membrane (2.50)
$P(\omega)$ Fourier spectrum of source signal (2.1)
P1, P2 Latencies at the peaks of auditory evoked potential
(P2 − P1) Latency-difference
Q^L Optimal subjective space with L-dimension (1.1, 2)
Q^M Auditory subjective space with M-dimension
r Integer
$r=(x, y, z)$ Receiving point (2.10)
R Gas constant (2.55, 56)
Re Real part
R_f Total reflection factor (2.39)
$r_0=(x_0, y_0, z_0)$ Position of a sound source (2.10)
R_r Scattering amplitude of rth spectral order (2.26, 6.1)
$r(x)$ Acoustic resistance of each section of basilar membrane (2.49)
$R(x)$ Membrane losses (2.51)
s Integer
S Number of subjects; total scale value of preference (5.9, 10, 19)
S_1, S_2, S_3, S_4 Scale values of preference as a function of listening level, initial time-delay gap, subsequent reverberation time and IACC, respectively (5.11, 13, 15, 17)
S' Scale value of preference for performer (6.15)
SD Scale of dimension of auditorium
$S(K)$ Imaginary part of Fresnel integral (2.18)
$S(\omega)$ Transfer function from a sound source to cochlea (2.48)
S_m Pressure spectrum, $m=1, 2, \ldots, M$ (A.4)
s_n $=n^2$, n^2 is taken as the least nonnegative remainder modulo N, and N is an odd prime (6.7)
$s(t)$ Single-pulse sound source (7.1)
$s(t)$ Signal received at a point r in a room (7.22)
$s(t)$ Impulse response of A-weighting filter corresponding to ear sensitivity (2.9)

$s(t)$	Impulse response for sound radiation of loudspeaker (4.10)
$s(x)$	Stiffness of each section of basilar membrane (2.49)
t	Time
T	Time interval (2.3); absolute temperature (2.55, 56)
T_m	Reverberation time of each comb filter (3.21)
T_{sub}	Decay rate to decrease to 60 dB after early reflections (3.21)
$[T_{sub}]_p$	Calculated preferred subsequent reverberation time ($\approx 23\,\tau_e$) (4.11, 5.8)
Δt_1	$=(d_1-d_0)/c$ (3.2, 5.21)
$\Delta t_1'$	Delay time of the single reflection to performers (6.15)
ΔT_i	Certain delay time for reproducing reverberation with "subjective diffuseness", $i=1, 2, 3$
Δt_n	Delay time of reflections relative to direct sound (3.2), $n=1, 2, 3, \ldots$
$[\Delta t_1]_p$	Calculated preferred initial time-delay gap between direct sound and first reflection (5.4)
$[\Delta t_1']_p$	Calculated preferred delay of the single reflection for performers (6.12)
$[\Delta t_2]_p$	Preferred second reflection relative to direct sound (4.8)
U	Velocity potential (2.10, 23)
U_i	Incident plane wave (2.26)
U_i	Velocity potential of each well (2.42)
$U_s(x;\omega)$	Reflection factor (6.3)
$V(x,\omega)$	Velocity of the basilar membrane (2.50)
$w_n(t)$	Impulse response describing reflection properties of boundaries (3.2), $n=1, 2, 3, \ldots$
$W(\omega)$	Transfer function describing reflection (2.16, 40)
$W_n(\omega)$	Transfer function for boundary reflections, $n=1, 2, 3, \ldots$
x	Cartesian coordinates
X_a, X_b	Discriminative processes of preference for any two sound fields A and B ($A, B \in F$)
x_r	Characteristic section of basilar membrane (2.53)
$x(t)$	Signal of the maximum length (7.11)
$x_1(t), x_2(t)$	Input signals to be applied to loudspeakers L_1 and L_2, respectively (3.9)
X^I	Optimal physically realizable space (1.1, 2)
X^N	Physically realizable space
y	Cartesian coordinates
$y(t)$	Output signal of the linear system (7.12)
z	Cartesian coordinates
$Z(x,\omega)$	Point impedance of basilar membrane (2.49)
$Z(x;\omega)$	Impedance at each well of diffusing ceiling (6.5)

$\alpha_1, \alpha_2, \alpha_3, \alpha_4, \alpha'$ Constants for weights of the scale value of preference (5.11, 13, 15, 17, 6.16)

α_0 $= k \sin\theta \sin\phi$

β_0 $= k \cos\theta$

γ_0 $= k \sin\theta \cos\phi$ (2.24)

$\delta(t)$ Dirac delta function (3.7, 7.5)

η Distance (2.11); elevation angle to a listener

η_n Specific acoustic admittance in a periodic structure

θ Elevation angle of incidence to a boundary wall

θ_i Incident angle (2.11)

θ_r Reflection angle (2.11)

κ^2 $= k^2 - \gamma_0^2$ (2.25)

λ Wavelength

λ_d Design wavelength ($= c/f_{low}$) (6.11)

ν Integer; time variable

ξ Distance (2.11); horizontal angle to a listener

π $= 3.14159\ldots$

ϱ Correlation coefficient between discriminative fluctuations in X_a and X_b (4.3)

σ Time delay; standard deviation (4.1)

σ_d Unit preference scale (4.3)

Σ Summation

τ Time variable, delay time

$\tau_{a,b}$ Delay time of all-pass filters (3.22)

τ_e Effective duration of autocorrelation function, defined by the delay at which the envelope of the normalized autocorrelation function becomes and then remains smaller than 0.1 (the ten percentile delay)

$\tau_{0,3}$ The thirty percentile delay of autocorrelation function

τ_m Delay time in comb filters (3.21), $m = 1, 2, \ldots, M$

τ_p Duration of autocorrelation function such that its envelope becomes $0.1A$, A being the total amplitude of reflections. This corresponds to the preferred initial time-delay gap (5.4)

ϕ Azimuth angle

$\Phi_{id}(\tau)$ Cross correlation between disturbing source and observation points (7.8)

$\Phi_{ii}(\tau)$ Autocorrelation function of input signal (7.4)

$\Phi_{io}(\tau)$ Cross correlation between input and output signals of an acoustic linear system (7.4)

$\Phi_{ll}(0)$ Autocorrelation functions at $\tau = 0$, corresponding to average intensities at left ear (7.30)

$\Phi_{lr}(\tau)$ Interaural cross correlation (3.5)

$\phi_{lr}^{(N)}(\tau)$ Normalized interaural cross correlation (3.7)
$\phi_p(\tau)$ Normalized autocorrelation function of a source signal (3.4)
$\Phi_p(\tau)$ Long-time autocorrelation function of a source signal (3.4)
$\Phi_{rr}(0)$ Autocorrelation functions at $\tau = 0$, corresponding to average intensities at right ear (7.30)
$\mathrm{Var}\{\Phi(\tau)\}$ Variance of autocorrelation function (2.8)
$\omega = 2\pi f$ Angular frequency
Ω_1, Ω_2 Lower and upper limits of audible frequencies [rad/s]
∇^2 div grad: Laplacian operator (2.23)
∂ Partial derivative

References

Chapter 1

1.1 M. R. Schroeder: Science **151**, 1355 – 1359 (1966)

General Reading

Békésy, G. von: *Sensory Inhibition* (Princeton University Press, Princeton 1967). It describes important phenomena and concepts in hearing as well as in other senses.
Jordan, V. L.: *Acoustical Design of Concert Halls and Theatres* (Applied Science, London 1980). This book describes practical experience in designing auditoria, collected during the author's long career as an acoustical consultant.
Kuttruff, H.: *Room Acoustics* (Applied Science, London 1979). This provides the fundamental of room acoustics.
Lee, Y. W.: *Statistical Theory of Communication* (Wiley, New York 1960). This supplies a fundamental background of the principles and techniques for analyzing such random phenomena as music and speech.
Roederer, J. G.: *Introduction to the Physics and Psychophysics of Music* (Springer, Berlin, Heidelberg, New York 1979). This suggests, in particular, how musical sound is detected by the ear and interpreted in the brain.
Schroeder, M. R.: *Number Theory in Science and Communication* (Springer, Berlin, Heidelberg, New York, Tokyo 1984). This supplies applications of number theory as a "queen of mathematics"; fast algorithms and how to apply the theory in concert hall acoustics.

Chapter 2

2.1 J. S. Bendat, A. G. Piersol: *Random data: Analysis and Measurement Procedures* (Wiley, New York 1971) Chap. 6
2.2 E. V. Jansson, J. Sunberg: Acustica **34**, 15 – 19 (1975)
2.3 A. N. Burd: Rundfunktechn. Mitteilungen **13**, 200 – 201 (1969)
2.4 Y. Ando: J. Acoust. Soc. Am. **62**, 1436 – 1441 (1977)
2.5 Y. Ando, K. Kageyama: Acustica **37**, 111 – 117 (1977)
2.6 M. Born, E. Wolf: *Principle of Optics* (Pergamon, Oxford 1964) Chap. 8
2.7 J. W. Rayleigh: *The Theory of Sound* (Macmillan, London 1878) Chap. 13
2.8 A. De Bruijn: Acustica **18**, 123 – 131 (1967)
2.9 A. De Bruijn: Acustica **24**, 75 – 84 (1971)
2.10 L. N. Deryugin: Dokl. Akad. Nauk USSR **87**, 913 – 916 (1952)
2.11 Y. Ando, K. Kato: Acustica **35**, 321 – 329 (1976)
2.12 Y. Ando, Y. Suzumura, K. Kato: Acustica **36**, 31 – 35 (1976/1977)
2.13 M. R. Schroeder: J. Acoust. Soc. Am. **65**, 958 – 963 (1979)
2.14 H. W. Strube: J. Acoust. Soc. Am. **67**, 453 – 459 (1980)
2.15 H. W. Strube: J. Acoust. Soc. Am. **70**, 633 – 635 (1981)
2.16 Y. Ando, G. Dai: Acustica **53**, 296 – 301 (1983)
2.17 Y. Onchi: J. Acoust. Soc. Am. **33**, 794 – 805 (1961)
2.18 S. Mehrgardt, V. Mellert: J. Acoust. Soc. Am. **61**, 1567 – 1576 (1977)

2.19 F. M. Wiener, D. A. Ross: J. Acoust. Soc. Am. **18**, 401 – 408 (1946)
2.20 E. A. G. Shaw: J. Acoust. Soc. Am. **56**, 1848 – 1861 (1974)
2.21 W. A. N. Dorland: *The American Illustrated Medical Dictionary*. 24th ed. (Saunders, Philadelphia 1947)
2.22 G. von Békésy: *Experiments in Hearing*, transl. and ed. by E. G. Wever (McGraw-Hill, New York 1960) Chap. 5
2.23 J. Tonndorf, S. M. Khanna: J. Acoust. Soc. Am. **52**, 1221 – 1233 (1972)
2.24 M. Rubinstein: J. Acoust. Soc. Am. **40**, 1420 – 1426 (1966)
2.25 E. H. Berger: J. Acoust. Soc. Am. **70**, 1635 – 1645 (1981)
2.26 J. J. Zwislocki: "The Acoustic Middle Ear Function", in *Acoustic Impedance and Admittance – The Measurement of Middle Ear Function*, ed. by A. S. Feldman and L. A. Wilber (Williams and Wilkins, Baltimore 1976)
2.27 A. T. Rasmussen: *Outlines of Neuro-Anatomy*, 3d ed. (William C. Brown, Dubuque, Ia 1943)
2.28 M. R. Schroeder: "Model of hearing", in Proc. of IEEE **63**, 1332 – 1352 (1975)
2.29 M. B. Lesser, D. A. Berkley: J. Fluid Mech. **51**, 497 – 512 (1972)
2.30 J. B. Allen, M. M. Sondhi: J. Acoust. Soc. Am. **66**, 123 – 132 (1979)
2.31 H. D. Patton: "Perception mechanism", in *Physiology and Biophysics*, ed. by T. C. Ruch and H. D. Patton (Saunders, Philadelphia 1965) Chap. 4
2.32 Y. Katsuki, T. Sumi, H. Uchiyama, T. Watanabe: J. Neurophysiol. **21**, 569 – 588 (1958)
2.33 G. von Békésy: *Sensory inhibition* (Princeton U. Press, Princeton 1967) Chap. 5
2.34 J. O. Picles: *An Introduction to the Physiology of Hearing* (Academic, London 1982) Chaps. 3 – 9

Chapter 3

3.1 Y. Ando: J. Acoust. Soc. Am. **62**, 1436 – 1441 (1977)
3.2 P. Damaske: Acustica **19**, 199 – 213 (1967/1968)
3.3 P. Damaske, Y. Ando: Acustica **27**, 232 – 238 (1972)
3.4 M. Barron: J. Sound Vib. **15**, 475 – 494 (1971)
3.5 V. Mellert: J. Acoust. Soc. Am. **51**, 1359 – 1361 (1972)
3.6 M. R. Schroeder, B. S. Atal: "Computer simulation of sound transmission in rooms", IEEE Intern. Conv. Rec. Part 7, 150 – 155 (1963)
3.7 Y. Ando, S. Shidara, Z. Maekawa, K. Kido: J. Acoust. Soc. Jpn. **29**, 151 – 159 (1973), in Japanese
3.8 M. Morimoto, Y. Ando: J. Acoust. Soc. Jpn. (E) **1**, 167 – 174 (1980)
3.9 M. R. Schroeder: J. Audio Eng. Soc. **10**, 219 – 223 (1962)

Chapter 4

4.1 L. L. Thurstone: Psychol. Rev. **34**, 273 – 289 (1927)
4.2 E. G. J. Eijkman: "Psychophysics", in *Handbook of Psychonomics*, Vol. 1, ed. by J. A. Michon, E. G. J. Eijkman and L. F. W. de Klerk (North-Holland, Amsterdam 1979) Chap. 6
4.3 W. S. Torgerson: *Theory and Methods of Scaling* (Wiley, New York 1958) Chap. 9
4.4 F. Mosteller: Psychometrika **16**, 207 – 218 (1951)
4.5 H. Gulliksen: Psychometrika **21**, 125 – 134 (1956)
4.6 H. Haas: Acustica **1**, 49 – 58 (1951)
4.7 Y. Ando, K. Kageyama: Acustica **37**, 111 – 117 (1977)
4.8 Y. Ando: "Preferred Delay and Level of Early Reflections in Concert Halls", Fortschritte der Akustik, DAGA '81, Berlin, 157 – 160 (1981), in which data by Y. Ando and Y. Shimizu are rearranged. Reports of the Meeting of the Acoustical Society of Japan, 621 – 622 (May, 1978), *see also* S. H. Kang, Y. Ando: "Comparison between Subjective Preference Judgments for Sound Fields by Different Nations", Memoirs of the Graduate School of Science and Technology, Kobe University, **3-A**, 71 – 76 (1985)

4.9 Y. Ando: "The Interference Pattern Method of Measuring the Complex Reflection Coefficient of Acoustic Materials at Oblique Incidence", Electronics and Communication in Japan **51-A**, 10 – 18 (1968), published by the Institute of Electrical and Electronics Engineers
4.10 Y. Ando: Appl. Acoust. **2**, 95 – 99 (1969)
4.11 U. Ingard, R. H. Bolt: J. Acoust. Soc. Am. **23**, 509 – 516 (1951)
4.12 G. von Békésy: *Sensory Inhibition* (Princeton U. Press, Princeton 1967) Chap. 5
4.13 Y. Ando, D. Gottlob: J. Acoust. Soc. Am. **65**, 524 – 527 (1979)
4.14 Y. Ando, M. Okura, K. Yuasa: Acustica **50**, 134 – 141 (1982)
4.15 Y. Ando, K. Morioka: J. Acoust. Soc. Jpn. **37**, 613 – 618 (1981), in Japanese
4.16 Y. Ando, K. Otera, Y. Hamana: J. Acoust. Soc. Jpn. **39**, 89 – 95 (1983), in Japanese
4.17 Y. Ando, M. Imamura: J. Sound Vib. **65**, 229 – 239 (1979)

Chapter 5

5.1 L. A. Jeffress: J. Comp. Physiol. Psychol. **41**, 35 – 39 (1948)
5.2 J. C. R. Licklider: Experientia **VII/4**, 128 – 134 (1951)
5.3 F. L. Wightman: J. Acoust. Soc. Am. **54**, 407 – 416 (1973)
5.4 J. A. Altman, L. J. Balonov, V. L. Deglin: Neuropsychologia **17**, 295 – 301 (1979)
5.5 Y. Kikuchi: "Hemispheric Differences of the Auditory Evoked Potentials in Normal Japanese", Audiology Japan **26**, 699 – 710 (1983), in Japanese
5.6 Y. Ando, I. Hosaka: J. Acoust. Soc. Am. **74**(S1), S64 – 5 (A) (1983)
5.7 R. Cohn: Science **172**, 599 – 601 (1971)
5.8 C. C. Wood, W. R. Goff, R. S. Day: Science **173**, 1248 – 1251 (1971)
5.9 J. L. Lauter: J. Acoust. Soc. Am. **74**, 1 – 17 (1983): See Fig. 2 in which data are cited from F. Spellacy's dissertation, University of Victoria, Canada (1968)
5.10 M. T. Wagner, R. Hannon: Brain and Language **13**, 379 – 388 (1981)
5.11 V. O. Knudsen: J. Acoust. Soc. Am. **1**, 56 – 82 (1929)
5.12 W. Kuhl: Acustica **4**, 618 – 634 (1954)
5.13 V. Fourdouiev: "Évaluation Objective de L'acoustique des Salles", Proceedings of 5th International Congress on Acoustics, Liège, 51 – 54 (1965)
5.14 T. Hontgast, H. J. M. Steeneken, R. Plomp: Acustica **46**, 60 – 81 (1980)
5.15 J.-D. Polack, H. Alrutz, M. R. Schroeder: Acustica **54**, 259 – 265 (1984)
5.16 K. Yamaguchi: J. Acoust. Soc. Am. **52**, 1271 – 1279 (1972)
5.17 Y. Ando: "Theory of Preference for Sound Fields in Concert Halls", Fortschritte der Akustik, FASE/DAGA '82, Göttingen, 182 – 186 (1982)
5.18 Y. Ando: J. Acoust. Soc. Am. **74**, 873 – 887 (1983)

Chapter 6

6.1 Y. Ando, T. Waki: "An Acoustic Improvement of a Hall Based on Subjective preference Judgments", Report of the Meeting of the Acoustical Society of Japan, 595 – 596 (October 1978), in Japanese, *see also* Y. Ando, D. Noson: J. Acoust. Soc. Am. **74**(S1), S83 (A) (1983)
6.2 M. R. Schroeder: J. Acoust. Soc. Am. **65**, 958 – 963 (1979)
6.3 Z. W. Qian: Personal communication
6.4 H. W. Strube: J. Acoust. Soc. Am. **67**, 453 – 459 (1980)
6.5 H. W. Strube: J. Acoust. Soc. Am. **70**, 633 – 635 (1981)
6.6 Y. Ando, M. Takaishi, K. Tada: J. Acoust. Soc. Am. **74**, 873 – 887 (1983)
6.7 G. M. Sessler, J. E. West: J. Acoust. Soc. Am. **36**, 1725 – 1732 (1964)
6.8 G. von Békésy: *Sensory Inhibition* (Princeton U. Press, Princeton 1967) Chap. 5
6.9 A. H. Marshall, D. Gottlob, H. Alrutz: J. Acoust. Soc. Am. **64**, 1437 – 1442 (1978)
6.10 I. Nakayama: Acustica **54**, 217 – 221 (1984)
6.11 I. Nakayama: "Preferred Direction of a Single Echo for Music Performers", Reports of the Meeting of the Acoustical Society of Japan, 487 – 488 (March 1984), in Japanese

6.12 K. Iida, Y. Ando, Z. Maekawa: "Effects of Stage Enclosure on Evaluating Sound Fields for Audience in Concert Halls", Reports of Meeting of the Acoustical Society of Japan, 543 – 544 (October 1984), in Japanese
6.13 L. L. Beranek: *Music, Acoustics and Architecture* (Wiley, New York 1962) Chap. 6
6.14 J. West: J. Acoust. Soc. Am. **40**, 1245 (1966)
6.15 A. H. Marshall: Archit. Sci. Rev. (Australia) **11**, 81 – 87 (1968)
6.16 A. H. Marshall: "Concert Hall Shapes for Minimum Masking of Lateral Reflection", Reports 6th International Congress on Acoustics, Tokyo, Paper E-2-3 (1968)
6.17 Y. Ando, Z. Maekawa: "Acoustic Design and the Objective Parameters of the Vega-Hall, Takarazuka", Architectural Acoustics and Noise Control, No. 41, 19 – 24 (1983), in Japanese
6.18 "Acoustics for all purposes – but mostly for muscic", Architectural Record, August 1984, pp. 130 – 132
6.19 Y. Ando, D. Noson: J. Acoust. Soc. Am. **74**(S1), S83(A) (1983)
6.20 P. H. Parkin, K. Morgan: J. Sound Vib. **2**, 74 – 85 (1965)

Chapter 7

7.1 H. Alrutz: "Ein neuer Algorithmus zur Auswertung von Messungen mit Pseudorauschsignalen", Fortschritte der Akustik, DAGA '81, Berlin, 525 – 528 (1981)
7.2 W. D. T. Davies: Control **10**, 302 – 433 (1966)
7.3 A. Lempel: Appl. Optics **18**, 4064 – 4065 (1979)
7.4 S. Osaki, Y. Ando: "Measurements of Objective Parameters of the Sound Fields using the Maximum-Length Sequence Signal", Joint Meeting of Societies Related to Electricity in Kansai Chapter, Himeji, Paper S12 – 2 (1984), in Japanese
7.5 M. R. Schroeder: J. Acoust. Soc. Am. **37**, 409 – 412 (1965)
7.6 M. R. Schroeder: J. Acoust. Soc. Am. **38**, 359 – 361 (1965)
7.7 M. R. Schroeder: J. Acoust. Soc. Am. **66**, 497 – 500 (1979)
7.8 R. Thiele: Acustica **3**, 291 – 302 (1953)
7.9 S. Osaki, Y. Ando: "A Fast Method of Analyzing the Acoustical Parameters for Sound Fields in Existing Auditoria", Proc. 4th Comp. Environ. Eng. Build., Tokyo, 441 – 445 (1983)
7.10 M. R. Schroeder, D. Gottlob, K. F. Siebrasse: J. Acoust. Soc. Am. **56**, 1955 – 1201 (1974); *see also* M. R. Schroeder: J. Acoust. Soc. Am. **68**, 22 – 28 (1980)
7.11 P. Damaske: J. Acoust. Soc. Am. **50**, 1109 – 1115 (1971)
7.12 P. Slater: Br. J. Stat. Psychol. **8**, 119 – 135 (1960)

Appendices

A.1 P. Damaske: Acustica **19**, 199 – 213 (1967/1968)
A.2 W. V. Keet: "The Influence of Early Lateral Reflections on the Spatial Impression", Reports of 6th Intern. Congress on Acoustics, Tokyo, Paper E-2-4 (1968)
A.3 M. Barron, A. H. Marshall: J. Sound Vib. **77**, 211 – 232 (1981)
A.4 Y. Ando, H. Sakimoto: "On the Relationship Between Spatial Impression and Calculated IACC for Sound Fields", Reports of the Technical Committee for Hearing and Architectural Acoustics for the Acoustical Society of Japan, Papers H-82-38 and AA-82-22 (1982), in Japanese
C.1 Y. Ando, H. Alrutz: J. Acoust. Soc. Am. **71**, 616 – 618 (1982)

Subject Index

J. G. Roederer

Introduction to the Physics and Psychophysics of Music

Heidelberg Science Library

Corrected Reprint of the 2nd edition. 1979.
79 figures, 7 tables. XIV, 202 pages
ISBN 3-540-90116-7

Contents: Music, Physics, and Psychophysics: An Interdisciplinary Approach. - Sound Vibrations, Pure Tones, and the Perception of Pitch. - Sound Waves, Acoustical Energy, and the Perception of Loudness. - Generation of Musical Sounds, Complex Tones, and the Perception of Tone Quality. - Superposition and Successions of Complex Tones and the Perception of Music. - Appendices: Some Quantitative Aspects of the Bowing Mechanism. Some Quantitative Aspects of Recent Central Pitch Processor Models. Some Remarks on Teaching Physics and Psychophysics of Music. - References. - Index.

This is the first basic textbook to establish the close relationship of physics, psychophysics, and neuropsychology to music.
The author analyzes the objective physical properties of sound patterns that are associated with the subjective psychological sensations of music. He describes how these sound patterns are actually generated in musical instruments, how they propagate through space, and how they are detected by the ear and interpreted by the brain. The approach throughout is scientific, but complicated mathematics has been avoided. The main revisions in the second edition reflect recent important developments in the understanding of complex pitch tone perception. Related developments in consonance and dissonance have also been incorporated. Other additions include a section describing the principal information channels in auditory pathways, and a section on the specialization of cerebral hemispheres in regard to speech and music. This reprint incorporates some corrections and slight changes.

Springer-Verlag
Berlin
Heidelberg
New York
Tokyo

Dynamics of the Singing Voice

By **Meribeth Bunch**

1982. 69 figures. XV, 156 pages
ISBN 3-211-81667-4

"This is among the best books of recent vintage that have crossed my desk and that I have been privileged to review. Written by a singer who is also trained in the biological sciences, its aim is to bridge the gap in communication and to foster understanding and collaboration between artists, teachers, therapists and doctors. It is not a 'How to' book, but rather one that addresses four major areas of concern: (1) the selection of a relevant terminology; (2) the formulation of an accurate description of the singing process; (3) the relating of existing knowledge of function to the teaching of voice; and (4) establishing the relationship of function to artistry.

Following the introductory first chapter, there is an excellent discussion of the psychological aspects of singing in which a number of key factors are shown to relate the voice to the whole person. In this section, the physiological aspects of singing are introduced, and these are expanded upon and clarified in subsequent chapters: 'Posture and Breathing in Singing'; 'Phonation'; Resonation and Vocal Quality'; 'Articulation'; 'Vocal Problems: Their Prevention and Care'; and 'Co-ordination, Spontaneity and Artistry'. The thrust of each chapter is to provide an understanding of the mechanisms of vocal production and the art of singing to aid in both rehabilitation and the prevention of injury and dysfunction.

The book is a model of organizational skill and writing craft. Its value is further enhanced by some 70 illustrations, with particularly good anatomical figures and an excellent bibliography of 539 items, providing a most comprehensive listing of the literature on singing. It is a dynaymic book; with it, Meribeth Bunch has given us a work of lasting value..." *NATS Bulletin*

Springer-Verlag
Wien

RETURN TO → **PHYSICS LIBRARY**
351 LeConte Hall 642-3122

LOAN PERIOD 1	2	3
1-MONTH		
4	5	6

ALL BOOKS MAY BE RECALLED AFTER 7 DAYS
Overdue books are subject to replacement bills

DUE AS STAMPED BELOW

MAR 2 0 1989	MAR 2 2 1994 APR 0 7 1994	APR 2 6 1996
March 28	Rec'd UCB PHYS	JUN 1 0 1996
April 24	MAY 1 6 1994	Rec'd UCB PHYS
DEC 2 1 1990	MAY 1 6 1994	MAR 2 6 1998
MAY 22 1992	Rec'd UCB PHYS	MAY 2 6 1998
Rec'd UCB PHYS	AUG 1 9 1995	MAY 1 5 2001
MAY 26 1992	AUG 1 1 1995	MAY 1 2 2002
JUN 26 1993 MAY 26 1993	Rec'd UCB PHYS	JUN 2 3 2003
Rec'd UCB PHYS	DEC 1 9 1995	
OCT 2 0 1993	AUG 1 1 1995	
Rec'd UCB PHYS	Rec'd UCB PHYS	
	FEB 1 7 1996 MAR 0 3 1996	

Rec'd UCB PHYS

FORM NO. DD 25

UNIVERSITY OF CALIFORNIA, BERKELEY
BERKELEY, CA 94720